GEOLOGICAL REPORT

OF THE

COUNTRY ALONG THE LINE

OF THE

SOUTH-WESTERN BRANCH

OF THE

PACIFIC RAILROAD,

STATE OF MISSOURI.

BY

G. C. SWALLOW,

STATE GEOLOGIST.

To which is prefixed a Memoir of the Pacific Railroad.

ST. LOUIS:

PRINTED BY GEORGE KNAPP & CO.

1859.

OFFICERS OF PACIFIC RAILROAD, 1858-9.

BOARD OF DIRECTORS.

JOHN M. WIMER,
JOHN J. ANDERSON,
H. I. BODLEY.
GEORGE KNAPP,
JOHN G. PRIEST,
H. CLAY HART,
JAMES McDONOUGH,
CHAS. P. CHOUTEAU,
SAMUEL GATY.
B. W. ALEXANDER,
BENJ. STICKNEY,
B. W. GROVER, of Johnson County,
GEO. R. SMITH, of Pettis County.

OFFICERS OF PACIFIC RAILROAD CO.

JOHN M. WIMER, *President,*
JOHN G. PRIEST, *Vice-President,*
FRED L. BILLON, *Sec'y and Treasurer,*
EDWARD MILLER, *Chief Engineer,*
DAN'L. TROWBRIDGE, *Auditor,*
T. McKISSOCK, *Superintendent,*
JOHN D. STEVENSON, *Agent of Land Department,*
THOMAS & COX, *Attorneys.*

HISTORICAL MEMOIR

OF THE

PACIFIC RAILROAD OF MISSOURI.

Previous to 1850, little or no attention had been given to the subject of internal improvements in the State of Missouri. A Board of Improvement had been created in 1840, but nothing further was done than to make a survey for a railroad from St. Louis to the Iron Mountain, by the way of Big river, and some surveys of the Osage river, with a view of improving its navigation.

The subject of a railroad across the continent having been discussed in various quarters, for several years, Col. Benton, then U. S. Senator for Missouri, on the 7th of February, 1849, introduced a bill into the United States Senate to provide for the location and construction of a Central National Road from the Pacific Ocean to the Mississippi river—to be an iron railway where practicable, and a wagon road were a railway was not practicable—and proposed to set apart seventy-five per cent. of the proceeds of the sales of the public lands in Oregon and California, and fifty per cent. of the proceeds of all other sales of the public lands, to defray the costs of its location and construction.

On the 20th February, a spirited public meeting was held at the Court-house in St. Louis, and a series of resolutions, introduced by Thomas Allen, was adopted, requesting the Legislature, then in session, to grant a charter and right of way, &c., for a railway across the State, from St. Louis to the western boundary.

On the 12th March, 1849, a charter was granted, providing for a capital of $10,000,000, and with "power to survey, mark, locate and construct a railroad from the city of St. Louis to the city of Jefferson; and thence to some point on the western line of Van Buren county, in this State, with a view that the same may be hereafter continued westwardly to the Pacific Ocean." The corporators named in the charter were John O'Fallon, Louis V. Bogy, James H. Lucas, Edward Walsh, George Collier, Thomas B. Hudson, Daniel D. Page, Henry M. Shreve, James E. Yeatman, John B. Sarpy, Wayman Crow, Joshua B. Brant, Thomas Allen, Robert Campbell, Pierre Chouteau, Jr., Henry Shaw, Bernard Pratte, Ernst Angelrodt, Adolphus Meier, Louis A. Benoist and Adam L. Mills.

In the spring of the same year another meeting was held in St. Louis for the purpose of calling a National Convention, to be held in St. Louis in October, and a committee of twenty-five citizens was appointed to make the necessary arrangements. A spirited address to the people of the United States, written by Thomas Allen, was issued, and a large convention, at which fifteen States were represented, of which the Hon. George Darsie, of Pennsylvania, was President, was held at the Court-house, in St. Louis, during the 15th, 16th, 17th and 18th days of October. This convention warmly commended the project of a National Pacific Railway across the continent, and made an address to the people of the United States and a memorial to Congress in its behalf.

In January, 1850, Mr. Thomas Allen, over his own signature, in the Missouri Republican, published the charter granted by the Legislature and called a meeting of the corporators. This meeting was held at the office of the St. Louis Insurance Company, on Thursday evening, January 31, 1850. There were present at this meeting, John O'Fallon, James H. Lucas, D. D. Page, Wayman Crow, Edward Walsh, George Collier, J. B. Brant, James E. Yeatman, Adolphus Meier, Adam L. Mills and Thomas Allen.

Mr. Allen made an elaborate address, which produced a decided impression, in favor of commencing the work of making railways in Missouri. At this time not a single railroad touched St. Louis on either side of the Mississippi, nor was any built in the vicinity. The Erie Railroad was not completed, and only 7,000 miles of railroad had been constructed in the United States.

The result of the meeting was an immediate organization ot a company, and a subscription on the spot, by the eleven gentlemen present, of $154,000. Books for further subscriptions were ordered to be opened; a committee was appointed to make arrangements for a general topographical and geological survey of the country, and to prepare a memorial to Congress for a donation of alternate sections of public lands along the route for the construction of the proposed road. At that time there were large bodies of public land in the State open to private entry, 29,216,000 acres, as was stated in the memorial to Congress, remaining unsold.

The preliminary meeting above alluded to was organized by the election of JOHN O'FALLON, President, THOMAS ALLEN, Secretary, and D. D. PAGE, Treasurer. This organization soon afterwards settled down into a more permanent form for the year 1850, in the selection of THOMAS ALLEN, President, Secretary and Treasurer, and JAMES H. LUCAS, Vice President. Early in the season, JAMES P. KIRKWOOD, of New York, was selected as Chief Engineer. Books of subscription having been opened at the Merchants' Exchange in St. Louis, the sum of one million of dollars was subscribed by citizens of St. Louis by the 1st March.

The following gentlemen were elected Directors for the first year: Thomas Allen, James H. Lucas, D. D. Page, Edward Walsh, George Collier, James E. Yeatman, L. M. Kennett, Louis A. Labeaume and James Harrison. The preliminary surveys were commenced on the 24th May, and were closed on the 29th November, 1850.

Five different lines were surveyed, embracing in the whole over 800 miles of survey. Taking into consideration as well the estimated cost of construction of the different lines, as the probable need of a branch to the Iron Mountain, and to the South-west part of the State, the location, by Mill creek valley, valley of the River Des Peres, and by the valley of the Meramec, was adopted on the 18th of June, 1851.

During the progress of the surveys the President personally

visited and addressed the people and the county courts of nearly every county from St. Louis to the western boundary, and also laid his plans before the Governor of the State, which the Governor, after due consideration, substantially adopted. The City and County of St. Louis, and the County of Jackson, subscribed to the stock. Petitions to Congress in behalf of a grant of land, as applied for by the Company, were circulated and numerously signed in all the counties along the proposed line, and in due time transmitted to Congress.

The first division of the road (33 miles) having been put under contract, the first spadefull of earth was removed, in the absence of the Governor, by the then Mayor of the City, Mr. Kennett, on the 4th of July, in the presence of a large and enthusiastic audience, who were first addressed by the President and Hon. Edward Bates. This memorable event took place at a point on the south bank of Chouteau Pond, on Mr. Minckes' ground, west of Fifteenth street.

At the ensuing session of Congress, held in 1850–'51, a bill passed the Senate of the U. S., granting for the railroad alternate sections of land for a space of six miles in width on each side, but was not reached in the House of Representatives. In the same winter of 1850–'51, the President having been elected to the State Senate, a plan for a complete system of railroads for the State was laid before the Legislature by him, including a form of State aid by a loan of the public credit. This plan, which was soon adopted with some variation of starting points, contemplated the issue of State bonds to the Railroad Company to an amount equal to the amount first to be advanced by the stockholders, the Company agreeing to pay the interest and principal of the bonds, and the State reserving a first lien on the road as security. The first act was approved February 22d, 1851, and provided for the issue to the extent of two millions of State bonds to the Pacific Railroad Company, in sums of $50,000, upon satisfactory evidence being furnished to the Governor, at each application, that a like sum of $50,000 had been expended by the Company, derived from sources other than State bonds; *and provided*, that the bonds should not be sold below par. These bonds having twenty years to run, and bearing six per cent. interest, were sold for a premium for more than a year and a half, and some were sold as high as 110. Some important amendments to the charter were granted at the same session, by an act approved March 1, 1851. Persevering in the effort for a grant of public lands, Congress, on the 10th June, 1852, passed an act granting to the State of Missouri the alternate sections of land in a strip six sections in width on each side of the line, for the construction of a railroad from St. Louis to the western boundary of the State. Soon after the passage of this act, the Company petitioned the Governor to call an extra session of the Legislature, and the then Governor, Mr. King, complied with the request. So largely had individuals entered the public lands the previous year or two, in consequence of the railroad surveys, that it was soon discovered that the grant would be of little value for constructing a railroad in a direct line westward from St. Louis to the Western boundary. Therefore, in view of the immense district of country lying at the South-west, known to be desirable in soil, climate and minerals, yet inaccessible, and also in view of the probability that a good route for the National road to California might be found along the 35th parallel, it was deemed advisable to make a fork in the line of road, and run the main trunk nearly west in the direction of Kansas, via the State capital, and the fork or Branch in the South-western direction. To the road from St. Louis to the point of divergence from the main line, and thence to the South-west boundary of the State, the State granted the lands by the act of December 20th, 1852, without bonus and with an exemption from taxation, until the road could pay a dividend; and with also a further loan of $1,000,000 to the main line, and $1,000,000 to the South-west branch. The right of preemption to actual settlers already on the lands, at $2.50 per acre, was, however, reserved.

Mr. Allen, President of the Company, was appointed the agent of the State to select the lands, and for that purpose he went to Washington City. The lands selected, and the schedule of which, as furnished by the General Land Office, has the force of a patent, amounted to about 1,200,000 acres.

The Pacific Railroad Company, having surveyed a route for a Branch Railroad to the Iron Mountain, to cross the Meramec near the mouth of Calvey creek, in Franklin county, and run on an interior ridge, west of Big river, via Potosi, and reported that the Iron Mountain could thus be reached by building about sixty miles additional of railroad, at a cost of two or two and a half millions, the Legislature granted a loan to the Company for that branch of $750,000. Some clamor springing up for a "direct line" to the Iron Mountain from St. Louis, this loan was subsequently yielded and transferred to the St. Louis and Iron Mountain Railroad. At the same session of the Legislature, a general railroad law was enacted, February 24, 1853, fixing the guage of all railroads in the State at five feet six inches, and providing for the levy and collection of taxes to pay subscriptions to railroads made by municipal corporations and for the conversion of the same into stock.

On the motion of the President of the Company, also, (then Mr. Allen), a loan was enacted for the first time, providing for a Geological survey of the State, and appropriating $10,000 a year for two years, for that object. One of the consequences which has flowed from that law is, the accompanying interesting report of the State Geologist, upon the geology of the country, rich in minerals, through which the South-west branch of the Pacific Railroad is located. An act was also passed the same day, authorizing the Pacific Railroad Company to extend, construct, and operate their road, or make contracts, &c., to any point west of the State boundary—having in view a connection with a National road.

It was on the 31st March, 1853, that Congress provided for such explorations and surveys as the War Department might deem advisable, in order to ascertain the most practicable and economical route for a railroad from the Mississippi river to the Pacific Ocean. Very soon after, Mr. Allen, being then in Washington, and believing a route in that direction practicable, proposed to the Secretary of War to make a survey, without expense to the government, for a railroad through South-western Missouri, and thence by the way of the Canadian river and Albuquerque, &c., to California (the route now, in part, travelled by the overland mail); provided, that the results of the survey should be reported with those of the Army Engineers. The proposition was declined.

In November, 1852, the first Locomotive, the "Pacific," manufactured at Taunton, Mass., was placed upon the track, at the machine shop erected by the Company, and ran out to the Manchester road, and was quite a noticeable and marked event in this State.

In December, 1852, a train, loaded with passengers, ran out to Cheltenham, about five miles, where a large company was entertained at a public dinner given by the President. During the same year, Thomas S. O'Sullivan, Esq., having been elected Chief Engineer, on the resignation of Mr. Kirkwood, the South-west branch was surveyed and located, under the immediate charge of his Assistant, James K. Ford, Esq.

In July, 1853, the First Division was opened from St. Louis to Franklin, a distance of 38 miles, and the event was appropriately celebrated at that place. In the spring of that year, the President being then in New York, a contract was made with Diven, Stancliff & Co., for the construction of the whole South-western branch. An ebb-tide in monetary affairs rendering it difficult to negotiate loans on Railroad securities, as contemplated, a meeting of stockholders was held at Mercantile Library Hall, and it was proposed that the County of St. Louis make an additional subscription of $1,200,000 to the capital stock of the Company, to be paid by taxation within four years, and that the stockholders claim no exemption from the tax, as the law allowed, in consequence of their being already subscribers. The President, Mr. Allen, who had devoted his time and energies to the starting of the enterprise, the first year without pay, and during the last at a salary of $1500 per annum, willing still to make sacrifices for the cause, and desirous of attracting public attention at once to the necessities of the case, and to propitiate all opposition, if any, on the score of long continuance in office, tendered his resignation; this was at length accepted, and Hudson E. Bridge, Esq., was elected President, April 25, 1854. At the election which took place, on the question of making the subscription and levying the heavy tax proposed on the part of the County of St. Louis, it was decided affirmatively by a vote of 3420 yeas to 1133 nays. Thus the people of St. Louis made themselves the first example in the

State under the law authorizing the imposition of taxes for railroad purposes.

The Legislature, by an act approved Feb. 10, 1855, modified the law in regard to the issue of State bonds as loans to the Railroads, so far as the amount was limited to the particular sum of $50,000 to be applied for at any one time, leaving the amounts within the general limit optional, and permitting sales at market price.

The work was pushed forward on all the contracts to Jefferson City that year; and in September, the work under the contract for the South-western road was also commenced.

A contract was also entered into to construct the main line of the road between Jefferson City and the Western boundary, with Messrs. Kirkwood, Porter & Co. The superstructure of the main line being principally laid nearly to Jefferson City, (125 miles,) a few of the bridges only requiring completion and trestle work being temporarily substituted, an excursion was planned, in view of the then approaching session of the Legislature, to take place on the 1st November. A splendid train of ten passenger cars started from St. Louis loaded with the *elite* gentlemen of the city and surrounding country, one lady only being present—as precious, goodly and cheerful a company as ever breathed the air of a hopeful morning—but was fatally precipitated, from too much speed and too weak a structure, at the foot of the first pier of the Gasconade trestle work, 108 miles from St. Louis. There the Chief Engineer, and thirty-one others, prominent and worthy men, were instantly killed. This lamentable accident caused some delay. The direct damage to the running stock of the Company was about $21,750.

It appearing, from experience, that the cost of our railroads had been under-estimated, it became necessary for the ensuing Legislature to make further provision for them. With wise precaution they first, on the 7th of December, 1855, enacted a law to secure the prompt payment of the interest on the State bonds, by requiring the setting apart out of the State Treasury, on the 1st July, 1856, the sum of $200,000 as a basis of a State Interest Fund, and the further sum of $100,000 annually for thirteen years. The Treasurer and Auditor of the State were made Commissioners of this Fund; and in case any Railroad Company failed to deposit with the Treasurer of the State a sufficient sum to meet the interest upon any State bonds loaned to them thirty days before the interest becomes due, the Auditor is required to pay out of the fund to the Treasurer a sum sufficient to pay the interest as it falls due, such sum to be refunded by the delinquent Railroad Company, under penalty of receiving no more bonds and forfeiting their road to the State.

The Legislature then, on the 10th December, 1855, enacted that the State bonds might be issued to the Railroad Companies in the proportion of two dollars of loan advanced for one expended by the stockholders, and thus granted the further sum of two millions to the main trunk line of the Pacific Road, and also transferred to said main line the one million before authorized for the South-west branch. The Company was also authorized to mortgage a million acres of their lands and Southwest branch, and issue their own bonds thereon to the extent of ten millions, to aid them to construct that branch, the State agreeing to guaranty three millions of those Company bonds, the proceeds to be expended on the first 114 miles of the South-west branch, reaching from Franklin to a point beyond the Gasconade river; but the Company were required to expend $50,000, to be derived from other sources, for every $100,000 of bonds to be guarantied. This act required the first division of that branch to be completed within three years from its date, under penalty of forfeiture of the road to the State, its lands and franchises, by operation of law, subject only to the mortgage above mentioned. That law also extended the privileges of actual settlers on railroad lands, by granting them rights of pre-emption at $2.50 per acre to the extent of fifteen miles from the road.

The act also created and established a Board of Public Works, consisting of three persons, not stockholders, to be (after the first appointed by the Governor) elected by the people for four years, the first election in 1856, and further required each Railroad Company to set aside and pay to the State Treasurer, every year, on State bonds thereafter to be issued, one and one quarter of one per cent. (1¼) on each 30 year bond, and two and one-half per cent. (2½) on each 20 year bond, sold or hypothecated. The Treasurer of the State and the Treasurer of each Railroad Company for the time being, were made Commissioners of the Sinking Fund thus created, and each Company was required to pay to the State Treasurer the semiannual interest on the bonds issued to them thirty days before the coupons should fall due. The State Treasurer was required to select one place in the city of New York for the payment of the interest on all the bonds issued by the State, and to give public notice thereof thirty days in advance. This was a good provision and is calculated to consolidate and strengthen the public credit, while it places the State upon its proper dignity in guarding its own honor. So long as the Companies continue to provide the means to pay the interest themselves, as they are required to do, the State Interest Fund and the Sinking Fund will accumulate, and suitable provisions are made for the safe and productive investment of the funds in securities readily convertible. Thus with the lapse of each year, with the increasing value of the railroads and their earnings, with the enhancement in the worth of the railroad lands, with the gradual accumulations of the Sinking Funds, and with the constant and steady increase of the public wealth, THE PUBLIC CREDIT OF MISSOURI, not now to be much more extended by the constitution, will be firmly established beyond all contingency.

James H. Lucas, Esq., was elected President of the Pacific Railroad Company in March, 1856, but resigned about a month afterwards, when William M. McPherson, Esq., was elected President in his place, and Mr. Edward Miller soon after was made Chief Engineer. Mr. McPherson continued to serve as President until March, 1858, when Hon. John M. Wimer was elected in his place.

By an act approved March 3, 1857, the State agreed to guaranty the bonds of the Pacific Railroad Company, issued as authorized by the act of Dec. 10, 1855, upon a mortgage of lands on the South-west branch in sums of $100,000 each, to an amount not exceeding $4,500,000—the first $100,000 to be issued upon evidence of a like amount of expenditure on that branch by the Company derived from sources other than guarantied bonds, but the subsequent amounts were to be issued as fast as each given sum was expended. The Governor was also authorized to make such guaranties in larger amounts than $100,000 at a time, if expedient, and place them for sale in the hands of an agent to be appointed by him, &c. The Company was required to complete the South-west branch in four years, pay the interest, and hold the State harmless from her guaranty, or forfeit the branch road, lands and franchises. The same act further provided, that whenever the Pacific Railroad Company had expended $500,000 west of Jefferson City, the Governor of the State should issue to them $1,000,000, part of the three millions granted by the act of Dec. 10, 1855, but not issued; and also granted a further loan of $300,000 of the same three million loan, to be based upon a showing of half that sum expended from stock subscriptions west of Jefferson City. And the act also granted the same Company a further loan of $1,000,000, to be issued in sums of $100,000, and the applications for them to be based upon proof of additional expenditure of half the amount derived from sources other than State bonds, and not included in any previous statement, and showing also that the proceeds of all the bonds issued under the act of 1855 had been expended in the construction of the road; and the statement of expenditure to be exclusive of interest, discount and commissions. These laws also provided, that the work should progress continuously west so as to leave no part unfinished beyond the reach of the means of the Company. The act also postponed the payments into the Sinking Fund, required by the act of 1855, until January 1st, 1859, when said payments are to commence and be made as before required, and within two years from that time the Companies are to make full payment of all sums thus postponed. It was the same act of March 3, 1857, which required the State Geologist to make a thorough survey along the lines of all railroads aided by the State, and to report in detail, to the President and Directors, "all the mineral, agricultural and other resources which may affect the value or income of the road under their direction."

By a proposed amendment to the constitution, which passed the Legislature by an almost unanimous vote, and approved March 4, 1857, the State debt is limited "to never exceed the sum of thirty millions of dollars." This will become a part of the constitution when passed upon again by the present Legislature. As affecting also the value of State bonds, as well as City and County bonds issued to railroads, and Railroad bonds also, all are exempted from taxation by the act of March 4, 1857. And, again, the Banks, under the act of 1857, are required to invest ten per cent. of their paid-in capital and two per cent. per annum of their nett profits in State bonds; and moreover, each Bank is required to pay for their privileges, annually, one per cent. on the amount of their paid-up capital, to the State, which is to go to the credit of the "State Interest Fund"—thus materially strengthening the credit of the bonds.

In consequence of the panic in the money market, the State bonds of Missouri, like many others, touched a low point in the fall of 1857, and many of the holders felt much alarmed. But, as was predicted by those best acquainted with the resources of Missouri, the Legislature met the crisis with a determined energy which inspired new confidence. The act of November 19, 1857, suspended the further issue and guarantee of bonds until March 1, 1859, with some exceptions; and among them $400,000 were permitted to be issued to the Pacific Railroad to finish to Round Hill, and $200,000 to carry the South-west branch to Moseley's. But it was agreed that whenever State bonds could be sold for 90 cents on the dollar, the Governor might issue $500,000 for the South-west branch, and receive in exchange the same amount of guarantied bonds, and to deposit with the State Treasurer a like amonnt of seven per cent. Railroad mortgage bonds as collateral security; and as the latter bore seven per cent. interest and the former six, the Company were required to pay the difference (one per cent.) into the State Interest Fund, on the bonds so exchanged. The Pacific Railroad was also required to deliver up all guarantied bonds; and a like amount of State bonds, running 20 years, and bearing six per cent. interest, were ordered to be issued and delivered to them. It was a singular fact that while State bonds sold readily, mortgage bonds, guarantied by the State, could not be sold. The act also authorized a special tax of one-tenth of one per cent. on the $100, to be levied upon all taxable property in the State, commencing in 1859, to be paid into the State Interest Fund; and also provided, that the per cent.

due from the United States to the State of Missouri, on account of the public land sales in the State, under the act of Congress of March 6, 1820, and of March 3, 1857, when received, should also be placed in the "State Interest Fund."

The Board of Public Works were required to attend all the meetings of the Boards of Directors, and watch their proceedings. And in order to provide further for the certain and prompt payment of the interest on any State bond which may be unprovided for, the Commissioners of the State Interest Fund, thirty days before the interest is due, shall temporarily take out of any funds in the public treasury, except the School Fund, the Road and Canal Fund, and the Internal Improvement Fund, sufficient to pay such interest; and in case there is not sufficient to pay such interest, then the Governor is authorized to issue "Revenue Bonds," payable two years after date, with any rate of interest not exceeding ten per cent., and hand them to the Commissioners, to sell or hypothecate for a loan in anticipation of the moneys due to the Interest Fund. And in case the moneys provided by the act are not needed to pay accruing interest, then they are to be invested, and the interest on the investment added to the fund, and the fund is declared sacred and inviolable until the principal and interest of all the State bonds are fully paid. Thus full and ample provision has been made by the Legislature to meet, at all times, the accruing interest on the State bonds of Missouri.

These facts are important as bearing upon the future of the Pacific Railroad in completing the line to Kansas, and the South-western branch.

The main, or Kansas line of the Pacific Railroad, was opened to Jefferson City (125 miles from St. Louis), about the 1st of March, 1856, and is well constructed and well stocked. Its business between that point, which is the State capital, and St. Louis, has exceeded the expectations and estimates of all. That line was opened to the town of California, 25 miles further, on May 4th, 1858, and to Tipton, 12$\frac{3}{4}$ miles still further west, being 162$\frac{3}{4}$ miles from St. Louis, July 25th, 1858. The amount expended upon that line to Dec. 1, 1858, has been about

$10,033,823, and for its construction $6,780,000 of State bonds have been issued to the Company.

The annual interest on these bonds is about - - - -	$406,800 00
The gross earnings of the road for the last twelve months were,	
From Freights, - - - - - - - - - -	$296,580 70
" Passengers, - - - - - - - -	320,791 44
" Mails, &c., - - - - - - - -	19,139 60
Total, - - - - - - - - -	$636,511 74

Upon the South-west branch about 19 miles of track are laid, and the next 60 miles can be completed in 1859. The iron for this distance is already contracted for at the Cambria Works in Pennsylvania.

The amount expended in construction upon the South-west branch, beyond Franklin, to Dec. 1st, 1858, is $1,442,710.

The amount of bonds issued for that branch to Dec. 1st, is $1,268,000 of State 6 per cents., and $132,000 guarantied 7 per cents.

The total length of the main line of the Pacific Railroad, as now definitely located from St. Louis to Kansas, is 282 miles.

The length of the South-west branch is 283 miles.

The total amount of State bonds issued to all Railroads is $19,056,000.

The total amount granted is $24,950,000, of which $5,894,000 are not yet issued.

The following statement of the issue and distribution of these bonds is believed to be correct:

Names of Roads.		*Received.*	*Remaining.*	*Total.*
Main Line Pacific Railroad		$6,780,000	$220,000	$7,000,000
South-west Branch	do.	1,400,000	3,100,000	4,500,000
North Missouri	do.	4,350,000	1,150,000	5,500,000
Iron Mountain	do.	3,276,000	324,000	3,600,000
Hannibal and St. Jo.	do.	3,000,000	....	3,000,000
Cairo and Fulton	do.	250,000	400,000	650,000
Platte County	do.	....	700,000	700,000
Total		$19,056,000	$5,894,000	$24,950,000

The security of the State, being the first lien, is based not only upon the Roads themselves and their appurtenances, upon which large amounts of private capital have been expended, amounting in the case of the Pacific Main Line alone to $3,254,582, but also upon large grants of land, amounting to over two millions of acres, of which the fee simple title has absolutely passed by act of Congress and the decision of the General Land Office.

INTRODUCTION.

Hon. JOHN M. WIMER,

President of the Pacific Railroad Company.

Sir:—In conformity to a law requiring me to make geological surveys along the lines of all the Railroads aided by the State, the country along the South-Western Branch of the Pacific Railroad, in which the lands of your company are located, have been examined with sufficient care to enable me to speak with certainty as to the general character of the country, the climate, soils, minerals, timber, and water power. In this survey we have extended our examinations over the counties of St. Louis, Jefferson, Franklin, Gasconade, Crawford, Phelps, Maries, Pulaski, La Clede, Webster, Green, Lawrence, Newton, and a part of Jasper, McDonald, Polk, Stone, Barry, Taney, Dallas, Washington and Wright—in all, an area of some 13,000 square miles.

In the time, and with the means allotted us for this work, it could not be expected that we would be able to speak with certainty of each section or even township in so wide an area; but by availing myself of the examinations previously made in portions of this region, and by so arranging the operations of the geological corps as to render the labors of each directly or indirectly available in this report, without materially retarding the progress of the State survey, we have been enabled to collect data sufficient to give a very correct view of the topographical, geological, mineral and agricultural features of the country under consideration.

I am indebted to Dr. Shumard for reports upon St. Louis, Jefferson, Franklin, Crawford, La Clede and Pulaski counties; to Dr. Litton for many valuable analyses of soils and minerals, and reports on numerous mines; to Mr. Price for the sketches of Granby and the Bluffs of the Niangua, for much valuable assistance in the field and in making up this report; and to Mr. Broadhead for a report on Maries county, for assistance in the field, in making up this report, and in preparing the accompanying section and map.

Mr. J. L. P. W. Fitzgerald, of Granby; Judge W. C. Price, Mr. J. A. Stephens, and Mr. Chas. Carlton, of Springfield; Mr. C. D. Bray, N. A. Davis, M.D., Mr. W. C. Smart, and Mr. L. P. Ayers, of Green county; Mr. C. L. Dickerman, Mr. D. G. Morrow, Mr. McCraw, Joel Hall, Esq., Mr. J. G. McFadden, Mr. Casswell Roberts, and Mr. Harvey Burkhart, of Taney; Hon. H. T. Blow, Mr. J. F. Darby, Mr. T. C. Johnson, and Mr. S. M. Colman, of St. Louis; Mr. Wm. C. Best, of Maries; Mr. Isaac N. Young, of Franklin; Pleasant Johnson, Esq., and John S. Reding, Esq., of Newton county; C. A. Edmands, of Washington; and Henry H. Fox, of McDonald, have rendered very important aid in our explorations and in collecting the statistics for our reports.

The large amount of material thus collected, has been carefully digested, and the most important results, such as are deemed entirely reliable, have been condensed into the following report. This report is submitted with a painful sense of its imperfections, and how far it comes short of fully representing the very extensive, interesting, and rich region upon which it is made. Still it is hoped we have presented such an amount of information as will very conclusively indicate the vast agricultural, mineral and manufacturing resources of that beautiful and favored country.

Wishing you eminent success in your able efforts to hasten the completion of the South-Western Branch,

I remain, very respectfully,

Your Ob't Servant,

G. C. SWALLOW,

State Geologist.

GEOLOGICAL ROOMS, STATE UNIVERSITY,
Columbia, Mo., June 2d, 1858.

GEOLOGICAL REPORT

OF THE COUNTRY ALONG THE PACIFIC RAILROAD AND THE SOUTH-WESTERN BRANCH, FROM ST. LOUIS TO THE WESTERN BOUNDARY OF THE STATE, IN NEWTON COUNTY.

It has been our object, in making this survey, to examine into all the available resources of this part of the State, and especially those designed to furnish a people with sustenance and wealth, and provide a surplus for trade and exportation; as a dense and wealthy population, and a surplus of productions, are the real elements of Railroad profits as well as national power and progress.

TOPOGRAPHY.

That portion of Southern Missouri extending from Newton county in the south-west, to Ste. Genevieve in the south-east, usually represented as the eastern extremity of the Ozark Mountains, is, in fact, a table land varying from 1,000 to 1,500 feet above the ocean. In the west it is sufficiently undulating to be well drained, while in the south and east it sometimes rises into ridges and knobs of moderate elevation.

From this table land, the country descends by moderate slopes in every direction. On the northern slope are the head waters of the Sac, Pomme de Terre, Niangua and Gasconade, flowing into the Missouri; on the east, the Meramec and the Big, flowing into the Mississippi; on the south, the waters of the St. Francois, the Current, and the White and its tributaries, descending towards Arkansas; and Spring River and Shoal Creek, on the western slope.

The valleys of the numerous streams which flow from this table land are at first but little depressed below the general level; but the farther they descend, the deeper and wider they become, until they expand into broad alluvial bottoms bounded by bluffs more or less precipitous.

This table land presents a surface sufficiently undulating to be well drained, and still level enough for argricultural purposes.

The South-Western Branch from Franklin, rises gradually onto the north-eastern slope of this table land, up the divide between the waters of the Bourbeuse and the Meramec, until it reaches an elevation of 780 feet above St. Louis, before crossing the valley of the Gasconade. Beyond this valley it rises again to the most elevated part of the line in Webster county. Thence through to Green, Lawrence and Newton, it descends the gentle western slope to the State boundary, where the road will be 440 feet above the St. Louis Register,* and the highest point on the line of the road at Marshfield is 1092 feet above the Régister, and about 1500 above the Gulf of Mexico. At Buck prairie it is 1020 feet above the Register, and 780 at Little prairie, east of the Gasconade.

CLIMATE.

This table land, as above stated, has an elevation of some twelve or fifteen hundred feet above the ocean. It has a rolling surface, and gentle slopes of some four or five feet to the mile, towards the valleys of the Osage, the Mississippi, the Arkansas, and the Neosho or Grand river, and no high mountains or arid plains to disturb the equable and agreeable temperature, which usually prevails at this altitude, under the thirty-seventh parallel of north latitude. There are no swamp or overflowed lands from which noxious exhalations can arise to affect any considerable portion of this country.

The climate, as these facts indicate and our meteorological observations clearly prove, is most agreeable and salubrious ;† the summers are long, temporate and dry, the winters short and mild. No climate, in short, is better fitted to secure health and a luxurious growth of the staple products of the temperate zone.

---

* This Register is about 400 feet above the Gulf of Mexico.

† The Census Report of 1850 shows this to be one of the most healthy regions in the United States.

GEOLOGY.

It is not deemed expedient, in this report, to enter into a detailed description of the rocks in the region under consideration, as that has been done in the *Second Annual Report of our Survey*, where any one desiring it can find a full exposition of these rocks, their classification, and catalogues of the fossils upon which that classification was based. In this connection, therefore, we shall merely mention the extent of each formation, its economical relations, and whatever may appear peculiar in the localities observed.

QUATERNARY SYSTEM.

All of the deposits of sand, clay, marl and humus in the bottoms of the streams, together with the clays and marls spread over the consolidated strata on the high lands, belong to the Quaternary period. These deposits cover the entire region, and are particularly important, as they furnish a large part of all the mineral ingredients that enter into the composition of the soils which rest upon them.

The Alluvial Formation, in the bottoms, is made up of clays, sands, marls and humus, more or less commingled. The character of these materials explains the wonderful fertility of the soils resting upon them.

The Bluff Formation consists of impure clays and marls, and is best developed on the eastern end of the line, particularly in St. Louis county, where it covers the high country and forms the basis of the soils. On the central and western portions, the superficial deposits are more argillaceous and sparingly developed, often leaving the underlying rocks to exert their due influence upon the soil. The Quaternary is represented on the map by the carmine color.

CARBONIFEROUS SYSTEM.

The Coal Measures underlie about 160 square miles in St. Louis county, a portion of St. Charles, and some small patches in Crawford, Phelps and Newton. It is represented by the purple color in the accompanying geological map. This formation in St. Louis county contains three beds of coal; one of hydraulic limestone and one or more of good fire-clay. The Coal Measures are represented on the map by the purple.

The Mountain Limestone underlies nearly all of Newton, Lawrence and Green, the southern part of Jasper, the north of McDonald, Barry and Stone, the south-west of Webster, and portions of St. Louis, Jefferson and St. Charles, and occupies an area of more than 3,000 square miles. The St. Louis, the Archimedes, and the Encrinital Limestones are developed in the East, the Ferruginous Sandstone, the Archimedes and Encrinital * Limestones in the West.

The Mountain Limestone contains the numerous and extensive deposits of lead and zinc in Jasper and Newton; the extensive beds of iron in Green and Lawrence; the marbles of St. Louis, and an abundance of good building stones in all the counties above named. It also exerts a good influence upon the soils, rendering them productive and durable.

It is represented on the map by the blue.

CHEMUNG GROUP.

This division is sparingly developed in St. Louis, Webster, Green, Taney, Stone, Lawrence, Newton and McDonald. Although this formation is very thin, its three divisions are well defined in some parts of the West, where the denudation of the middle clay beds has formed the mounds so conspicuous in the prairies of that part of the State.

These sandy clays often exert an injurious influence upon the soil. No valuable minerals have been found in these rocks, save some copper in the beds of transition between them and the Encrinital Limestone in Lawrence county. It occupies an area of some 160 square miles, and is represented by green on the map.

* This formation presents some peculiar features in Lawrence county. Its lower part is there made up of heavy beds of whitish, porous quartz rock, as indicated by the following section on the Turnback:—

No. 1—5 feet of Ferruginous Sandstone.
" 2—30 feet Encrinital Limestone, with its usual characters.
" 3—70 feet of brownish gray, porous and hard silicious rock or quartzite.
" 4—5 feet of soft brown impure Sandstone, with masses of calcareous spar.
" 5—10 feet of coarse, impure crystalline Limestone; contains *Spirifer Marionensis*—Chemung.
" 6—20 feet of silicious rock like No. 3.

LOWER SILURIAN.

The Trenton Limestone underlies a portion of St. Louis, Franklin and Jefferson counties. No valuable ores have as yet been discovered in this formation. It furnishes good limestones and marbles for building and ornamental purposes. The soils formed from it are calcareous and durable.

The Magnesian Limestone Series * occupies a large part of Jefferson, Franklin, and Webster, and nearly or quite all of Gasconade, Crawford, Washington, Maries, Phelps, Pulaski, La Clede, Wright, and Dallas—an area of nearly 6,000 square miles within the limits of the Railroad lands.

This series of ancient deposits is made up of magnesian limestones, sandstones and intercallated beds of chert or impure flint. These rocks contain the best building material in the State. Some of the limestones are not surpassed in beauty and durability. Many of them will furnish inexhaustible supplies of beautiful, variegated marbles. A few of the sandy beds are excellent free stone well adapted to architectural purposes; while others will afford any desirable quantity of pure white sand for cements and glass manufactures.

These rocks usually exert a good influence upon the soil; but there are exceptions. Some of the magnesian limestones decompose so rapidly, and supply the soil with so large a portion of magnesia as to impair its fertility; as is evident in many of the glades on the ridges and slopes, where they come to the surface. In a few localities the sandstones render the soil too arenaceous, while in other places the fragments of chert are so abundant as to prevent its use for ordinary cultivation. It should be borne in mind, however, that these apparent defects, when not in very great excess, give the soil a peculiar adaptation to one of the most important departments of husbandry, the culture of the grape. And the time is not far distant when

---

* The upper part of this series belongs unequivocally to the age of the Calciferous Sandrock; but since our Second Annual Report was published, the evidence that the lower beds are Potsdam Sandstone has been increased by the discovery of the *Lingula antiqua* of Hall, by Mr. Broadhead, in Moniteau county

the "*poor flint ridges*" and terraced slopes * of Southern Missouri, will be more valuable for vineyards than the best lands of the State for the other departments of agriculture.

This is, emphatically, the *mineral bearing* rock of Missouri. It contains the larger part of all the lead, zinc, copper, cobalt, and nickel, and a considerable portion of the iron discovered in the State. Some or all of these ores abound wherever these rocks have been explored within the limits of the State.

When it is borne in mind, that they occupy an area of some 10,000 square miles in the counties containing Railroad lands, we shall be less surprised at the long catalogue of mineral localities in those counties already known and be better prepared to expect still other discoveries of equal importance.

The Lower Silurian Strata are represented by yellow on the accompanying geological map.

IGNEOUS ROCKS.

There are but few unimportant representations of this division in the region under consideration. At one locality in Laclede, and one or two in Crawford, granite dykes, or ridges, rise above the stratified rocks.

SOIL.

The soils of this region, are as diversified and varied as the topographical and geological features, already disclosed, would indicate. The wide diffusion of the rich silicious marls of the Bluff formation, particularly in the eastern and western counties, is a sufficient proof of the value of the soils found upon it.

It is well known that a part of St. Louis and the adjoining counties possess a very superior soil. It may not be as well known, though equally true, that portions of Newton, Green, Lawrence, and the adjoining counties, have a soil equally good. It is also true, contrary to the opinions of some, that the central counties on the line of this road, have large areas of most excellent land.

---

* See Plate VIII.

Almost every acre of the alluvial bottoms throughout this entire region, has a rich, durable soil, which is usually well adapted to the culture of corn, wheat, tobacco, oats, and the grasses; some would yield good hemp. Where the silicious marls of the Bluff are well developed, the upland soils are rich, fertile and durable. This variety of soil prevails in all the best upland on the line of the road, particularly in the eastern and western extremities. In Oliver's prairie, Pool's prairie, and Sarcoxie prairie, in Newton; Grand and Kickapoo prairies, in Green; Pleasant prairie, in Webster; Dimond prairie, in Jasper; and Ozark prairie, in Lawrence, the soil is excellent. It possesses the same good qualities in some of the timbered portions of all the counties above named; but St. Louis county has much the largest proportion, as indicated by the superior soils in the valley west of the city and in the Florisant.

There is a soil somewhat inferior to the preceding, which covers large areas in the region under consideration. It also rests upon the marls of the Bluff where that formation is somewhat clayey and where it has been injured by washing. This variety is found on the ridges and undulating portions of the country, where the white, post and black oaks, and summer grapes abound, and white hickory, dwarf sumac, and hazle are less prevalent. This same soil also occupies the prairies, which are somewhat inferior to those mentioned above.

The following analyses show the qualities of this variety of soil:

Analyses of Soil from the Bluffs of Boone County,

BY DR. LITTON.

	No. 12 A*	No. 12 C*	No. 12 B*
Water expelled by drying at 150° C	0·4105	0·8030	0·6558
Organic matter & water not expelled at 150°C	3·0957	3·8901	2·6049
Silica, etc., insoluble in Hydrochloric acid ..	90·1420	85·0571	90·8063
Soluble silica	0·1384	0·2187	0·1475
Alumina	3·0654	4·7672	2·9346
Peroxide of iron	2·0553	3·8814	2·0590
Oxide of magnesia	a trace.	a trace.	a trace.
Lime	0·2086	0·4722	0·1242
Magnesia	0·3423	0·6581	0·2048
Potash	6·3368	0·3895	0·2121
Soda	0·1828	0·1220	0·2925
Phosphoric Acid	0·0560	0·0556	0·0346
Sulphuric Acid	0·0035	0·0099	0·0508
Chlorine	0·0000	0·0276	0·0000
Total	100·0373	100·3524	100·1311

This soil covers several thousand square miles in the counties comprised in this survey. It is the very best soil for wheat, and rye, in the State. It is well adapted to corn, tobacco, oats, and grasses; and is very much improved by deep cultivation, as the above analyses show the richest portions to be ten or twelve inches below the surface. The vineyards of Boonville, Hermann, and Hamburg, are on soils similar to this; and it produces most excellent wild grapes.

But the soils derived from the magnesian limestone series, cover the largest portion of this region. The sand, lime, magnesia and alumina, derived from the decomposition of these rocks, together with the abundance of vegetable matter from the decay of the rank vegetation, and the alkalies from the fires which annually overrun this country, combine to form a soil *light*, *dry*, *warm*, and rich in *potash*, *soda*, *lime*, *magnesia*, and all the other ingredients needed to render it fertile and

* No. 12 A was collected from 2 to 6 inches below the surface; No. 12 B, from 10 to 12, and No. 12 C, from 18 to 20 inches below the surface, on a high ridge covered with white, post and black oak, white hickory, dwarf sumac, hazle and summer grapes.

suitable in an eminent degree for many of our staple crops, and especially for

GRAPE CULTURE.*

Notwithstanding the true principles of grape culture are so little understood by the community at large, no department of agriculture has been more carefully investigated, more distinctly defined and reduced to scientific principles. Since Virgil wrote his masterly treatise upon the habits and cultivation of the vine, the principles which should govern its culture, have been within the reach of all who would investigate the structure of this plant and learn the soil and climate adapted to its perfect development. And, indeed, it could scarcely be otherwise, as the vine has occupied so prominent a position in the husbandry of almost all the enlightened nations of ancient and modern times.

Since Noah planted a vineyard, the vine has followed the progress of husbandry and civilization throughout India, Arabia, Palestine, and Southern Europe. It holds an important place in the history of those seats of ancient civilization and progress. The "vine-clad hill" occupied a conspicuous position in every landscape, and the juice of the grape had its place at the social board and ruled the joys of the banquet hall. While it held so important a position among the nations, its value led the ablest minds to investigate its habits and deduce the best modes of culture from the experience of the many engaged in the pleasant pursuit. Solomon investigated the properties of the vine, and Virgil gave so excellent a treatise upon its habits and culture that the investigations and experience of the last two thousand years have added but little to the knowledge then possessed.

Since then the habits of the vine, and the modes of culture best adapted to it, have been so carefully determined, and so thoroughly established by the experience of the last four thousand years, it only remains for the cultivators of our times to investigate the modes of culture so long and so successfully

* The vast importance of this subject, induced me to make a most thorough examination of all the facts showing the adaptation of the soil and climate of this region to the culture of the grape. The results of these investigations are most satisfactory.

practised in India and the countries bordering upon the Mediterranean; to inquire how far the varieties there cultivated, and the culture there adopted, will succeed in other localities; to determine whether some new varieties may not succeed better in other climates and soils; and what modifications of culture will secure the highest degree of success in the various soils and climates to which we would introduce the vine.

It is obvious that the success of the grape depends upon the mutual adaptation of both soil and climate. In places where the soil has all the requisite properties, the climate may be such as to prevent full success; as in many parts of New England, where the climate is too cold; and in England, where it is too moist. In many localities in Southern Europe, the soil is such as to prevent the full success of the vine, though the climate is all that could be desired.

Soil.—According to Virgil* and the best authors who have followed him, the soil should be *warm*, *light*, *dry*, and *rich in alkalies and alkaline earths*, especially *potash*, *soda*, *lime* and *magnesia*. The best vines have been grown† upon soils of this description; and when any of these qualities have been wanting, the most skillful vine-growers have supplied the deficiency by artificial means. Hence Virgil directs to place "porous stones and rough shells" in the trenches—the stones and shells to loosen the soil and perfect the drainage, the shells to supply the deficit of lime.

The vine has ever succeeded the best, other things being equal, in a calcareous soil. The best vineyards upon the Rhine, the Ohio, and the Missouri, are upon soils rich in lime; and, according to D'Orbigny, the wines from such vineyards in France are "more lively and spirituous."

The chemical composition of a plant also gives us sure indications of the mineral ingredients of the soil required for its perfect development. The following table, from Johnston's Agricultural Chemistry, contains the compositions of five vines, grown on five different soils. The result shows most conclu-

---

* Geor. Lib. II., lines 217—221 and 262.—"*Optima putri arva solo.*"

† The great vine at Windsor Park was planted fifty years ago. "In 1850," says Prof. Lindley, "it produced 2,000 large bunches of magnificent grapes, filled a house 138 feet long and 16 feet wide, and had a stem two feet nine inches in circumference. The border in which it grows is *warm, light, dry* and *shallow.*"

sively what mineral substances are demanded for the perfection of the vine :

	By Leibfrauen.	By Weincheimer.	Primary Rocks. Gratz.	Mountain Limestone. Gratz.	Mica Slate. Gratz.	Mean.
Potash	17·32	25·24	34·13	24·93	26·41	25·60
Soda	28·50	2·74	8 03	7·31	8·79	11·07
Lime	29·75	40·75	32·67	37·59	33·47	34·85
Magnesia	9·78	7·49	4·66	7·12	9·16	7·64
Oxide of Iron	4·12	1·52	0·16	0·24	0·19	1·25
Phosphoric Acid	5·20	18·89	16·35	19·55	16·87	15·37
Sulphuric Acid	1·96	2·88	2·16	2·37	2·44	2·36
Chlorine	1·82	0·53	0·50	0·35	0·25	0·68
Silica	1·55	....	1·45	0·62	2·48	1·22
Total	100·	100·	100·11	100·08	100·06	100·04
Percentage of Ash in dry twigs	2·835	2·689	2·525	2·25	2·325	2·525

These analyses show that *potash*, *soda*, *lime*, *magnesia* and *phosphoric acid*, enter largely into the composition of the vine, and that grapes will succeed best on soils rich in those materials. The other ingredients are such as are found in nearly all soils and may be left out of our investigations.

It is a well established principle of vegetable science that *lime* may supply the place of *soda* and *potash*, in part at least, in some plants. The following analyses of vines from two localities show this to be true of the vine also :

	I.	II.
Alkalies	45.82	27.98
Lime	29.95	40.75

If, therefore, soda and potash be deficient in a soil, their places may be partially supplied by lime, should it exist in sufficient quantities.

Climate.—The success of the grape on the islands and the shores of the Mediterranean, shows their adaptation to a climate in which the winters are short and mild, and the summers are temperate and equable. In the Ionian Islands, where the grape attains great perfection, it is never exposed to pinching cold or burning heat, or to any very sudden changes from one to the other. But the great profusion and excellence of the grapes in India, at Candahar and Cabul, " the sunny home of the grape," indicate an ability to reach perfection in spite of sudden changes from extreme cold to burning heat. "In no part of the world," says Lindley, "are the grapes more delicious than in Candahar and Cabul;" and yet the traveller speaks of the "*bitter cold wind* and *blazing fires* at *night*," and "the *burning sun* by day" in March, and the sun's heat at 140° in May, where the grapes ripen as early as June.

We may conclude then that the grape will, under favorable circumstances, reach the greatest perfection, though exposed to sudden changes and extremes of heat and cold.

Having ascertained the conditions of soil and climate best adapted to the successful culture of the vine, it has been my aim, during the progress of the Geological Survey of Missouri, to determine how far these conditions are fulfilled in Missouri; to what extent and with what succes the vine may be cultivated in our State, and the advantages to be derived from its cultivation.

In order to secure the most accurate data for our conclusions, our investigations have been directed to the following subjects:

1. The characters and habits of all our native vines, and the soils on which they succeed best, have been carefully noted.

2. Five persons* have been appointed to make meteorological observations; one at Springfield in the south-west, one at Cape Girardeau in the south-east, one at Palmyra in the north-east, one at St. Joseph in the north-west, and one at Columbia in the centre, in the valley of the Missouri river. These observers have been supplied with the very best instruments, and they have made and recorded their observations according to the plan adopted by the Smithsonian Institute.

3. The experience of our most successful vine-growers, has been collected, and the results carefully compared with the

---

* It gives me great pleasure to bear testimony to the disinterested labors of those who have so faithfully observed and recorded the meteorological phenomena at the stations above named. Our State will be under many obligations to the Rev. G. P. Comings, of St. Paul's College, Palmyra; Rev. James Knoud, of St. Vincent's College, Cape Girardeau; J. A. Stephens, Esq., of Springfield; E. B. Neely, A.M., of the St. Joseph High School; and Miss M. B. Hill, at Columbia, who have made the observations at their several localities.

conclusions derived from our examinations of the climate, soils, and wild vines of the State.

4. The soils of the State have been carefully observed, and the varieties collected and submitted to a most skillful chemist for full and accurate analyses.

Native Grapes.—The growth and fruit of our native vines give us most important indications of the adaptation of our soil and climate to the cultivation of the grape. The following species have been observed, and the growth, habits, and fruit of each variety, have been carefully examined.

1. VITIS LABRUSCA, *Linn.* *Fox Grape* of the Northern States.

This vine is abundant in all parts of the State. It attains to a very large size* in our rich alluvial bottoms and on our best upland soils; but the vines of a smaller size, which grow upon the dry ridges, on the declivities of the bluffs (especially those of the Magnesian Limestone), and on the talus of debris at their bases, exhibit a healthy, firm growth, and produce an abundance of fine fruit. The grapes found in these localities are larger and the pulp is more juicy and palatable.

Many well known and excellent varieties of grapes now in cultivation were derived from this species. The *Isabella*, *Catawba*, *Schuylkill*, and *Bland's*, are the most esteemed.

2. VITIS ÆSTIVALIS, *Michx.* *Summer Grape.*

This, like the preceding, is found in all parts of the State, and is doubtless the largest of all our vines. It is one of the most striking objects in our magnificent forests. While the stem, like a huge cable, hangs suspended from the limbs of the largest trees, the branches clothed in rich foliage, and often loaded with fruit, hang in graceful festoons over the highest boughs. But the vines growing on the thin soils of our limestone ridges and bluffs, and on the loose debris at their bases, where they are more exposed to the air and the sun, produce a greater abundance of the very best fruit.

* This vine often attains to a diameter of 10 inches, ascends the loftiest trees and spreads its branches over their highest boughs.

3. VITIS CORDIFOLIA, *Michx.* *Winter* or *Frost Grape.*

This vine is widely diffused through the State; but it is not so large as the Fox or the Summer Grape. Its fruit is small and acerb.

4. VITIS RIPARIA, *Michx.* *River Grape.*

This grape is partial to the alluvial soils along the margins of our streams. It grows to a large size.

5. VITIS VULPINA, *Linn.* { *Muscadine* of the West, and *Fox Grape*, according to Elliott, of the South-eastern States.

It is most abundant in the southern part of the State. It grows very large and produces abundantly. Its fruit is very much esteemed. The cultivated *Scuppernong Grape* is a variety from this species.

6. VITIS BIPINNATA, *Michx.*

This plant was observed in Cape Girardeau and Pemiscot counties.

7. VITIS INDIVISA, *Willd.*

This vine abounds in the central and western counties.

From this list it will be seen that Missouri possesses all the native grapes of our country save one, the *Vitis Caribœa?* (D. C.) of California. The vines are so abundant and so large as to form an important and conspicuous part in every copse and thicket throughout the entire State. They are everywhere present, lending grace and beauty to every landscape, and indicating with prophetic certainty that the day is not far distant when the purple vineyards will cover our hills, and the song of the vine-dresser will fill the land with joy, and the generous juice of the grape will improve our moral, intellectual and physical powers.

*Experience of our Vine-dressers.**—Several vine-dressers in

---

* I am indebted to Mr. William Haas, of Boonville, Mr. George Husmann, of Hermann, Mr. Frederic Mench, of Marthasville, and Mr. Joseph Stuby, of Hamburg, for valuable information respecting the cultivation of grapes in our State.

our State have been engaged in the cultivation of the grape during the last twelve or fourteen years. Their success has been fully equal to their expectations, and they are full of high hopes of the most useful and profitable results, even of *entire* and *permanent success.* Their experience in cultivating the vine has led them to the same conclusion that we have deduced from our scientific examinations of the soil, climate and native vines, viz: *that the vine can be cultivated with entire success in favorable localities in all parts of the State.*

It should be borne in mind that these results have been derived mostly from vineyards in the valleys of the Missouri and Mississippi rivers, which are not, by far, the most favorable localities in the State; for the "mildew" and the "rot," the most formidable obstacles they have had to contend with, may be partially or entirely obviated in localities where the atmosphere and soil are not so densely charged with moisture.

"The rot," says one of our most successful vine-dressers, Mr. Haas, "attacks the berries when the soil is in a wet condition in July and August." "It is most severe on the low and wet parts of the vineyard."

Mr. Husmann says: "The principal cause, all are agreed, is an excess of moisture about the roots, and damp, moist weather."

Now the larger part of our vineyards are located upon a *stiff, cold, clayey subsoil*, which, of necessity, retains the excess of moisture, and produces the injurious results.* This evil may be obviated by thorough draining, or, what is better, by selecting some of the millions of acres in the southern part of the State, where the soil is *warmer* and *lighter* and *richer* in the ingredients most favorable to the vine, and where the subsoil is so porous as to permit a free passage to the excess of moisture.

The mildew appears in June, and all agree that it is caused by "*foggy, damp* and *hot weather after rains.*" Now our observations prove that hot damp weather, accompanied by mists, is much more prevalent in the valleys of the Missouri and the Mississippi than on the table lands to the south.

The characters of the two regions under comparison, show

---

* See Soil No. 12, page 10.

most conclusively that the excess of moisture in the valleys must be considerable and permanent. The valleys are covered with numerous and extensive lakes, sloughs, and forests of rank growth and vast extent, besides the broad rivers which flow through them; while the table lands are almost destitute of lakes and ponds, and but partially covered by a very sparse and much less vigorous growth of timber; and besides, they occupy an elevation several hundred feet above the valleys.

No fears, therefore, need be entertained that these obstacles will prevent the entire success of vine culture in Missouri, should our atmosphere even continue as moist as at present. But we may expect much improvement in this respect, as it is fully established by past experience, that the settlement of a country, and the opening of a soil to cultivation, lessen the amount of rain and moisture in the atmosphere.

Notwithstanding the many difficulties our vine-dressers have had to contend with, and notwithstanding some of their vineyards are not, to say the least, in the most favorable localities in the State, their success has been very flattering.

The vineyards of Boonville have yielded, the present season, about 6,000 gallons, worth $12,000. Five acres gave a clear profit of $2,000, or $400 per acre. Mr. Haas made 1,550 gallons from three acres.

The vintage of Hermann was about 100,000 gallons, from less than 200 acres. At $1.00 per gallon—which is much less than the value—it will give a profit of at least $400 per acre, or of $80,000 on the 200 acres in cultivation..

One small vineyard at Hamburg, Mr. Joseph Stuby's, yielded over 1,000 gallons per acre.

The entire cost of vineyards, preparing the soil, setting and training the vines until they come into bearing, varies from $200 to $300 per acre.

Annual cost of cultivation after··$50 to $60 per acre;
Ten per cent. on first cost········$20 to $30 per acre;
Total expense for each year·····$70 to $90 per acre;

so that an income of $100 per annum for each acre is sufficient to pay the interest on the first cost and the expense of cultivation.

Judging from the statistics before me, I would suppose all our vineyards have yielded an average of at least 250 gallons per acre since 1849, which, at an average price per gallon of $1.60, would give an annual income of $400, and a yearly profit of $300 per acre. So that the vine-dresser, even in the poorest seasons, can scarcely fail of a handsome profit, while in good years his gains will far surpass those derived from any other department of husbandry. But the profits of our most successful cultivators have been much greater. Mr. Pœschel, of Hermann, is said to have made over 400 gallons per acre for the last ten years, and an annual profit of more than $500 for each acre.

Such are the favorable results legitimately derived from the experience of our vine-dressers in their early efforts in a new country, with a soil and climate unknown to the cultivators of the grape. All must admit that they are most satisfactory. Even if our climate does not become more dry, if no more improvements are made in the modes of culture, and if no more favorable localities are obtained, grape culture must increase very rapidly, and become an important element in our agricultural and commercial interests.

Soil.—Nearly all the soils of Missouri possess all the ingredients necessary to the complete development of the vine; but some of them are too *heavy*, *wet* and *cold*, unless improved by artificial means. This is true to some extent of those on the bluffs of the Mississippi and Missouri, where nearly all the vineyards of our State are located. Still, they produce an abundance of large native grapes, on vines of the *Vitis labrusca*, and other species.

The character of this variety of soil is indicated by the analysis of a specimen from the bluff of Boone county, as given above. It has already been shown that it covers large areas in the region under consideration. The superior native grapes growing upon this soil, and the success of the vineyards above named, prove its adaptation to the vine. Its greatest defect is a capacity to hold and retain an excess of water, which must be remedied by trenching and a proper admixture of vegetable matter, sand, pebbles and broken rocks.

But the action of the elements upon the rocks of the *Magnesian Limestone Series*, has prepared a soil, as if by design, to invite the vine-dresser to possess and cultivate it.

The following analysis shows the properties of this variety of soil:

ANALYSIS OF A MAGNESIAN LIMESTONE SOIL FROM THE SOUTHERN BLUFFS OF CALLAWAY COUNTY, BY DR. LITTON.

Soil No. 14.

Water expelled by heating to 150° C	1.1700
Organic matter and water not driven off at 150° C	9.6299
Silica, etc., insoluble in hydrochloric acid	54.2600
Soluble silica	0.1639
Alumina	10.8588
Peroxide of iron	2.5186
Manganese	a trace
Lime	8.0720
Magnesia	1.6609
Potassa	1.6378
Soda	0.3442
Carbonic acid	10.1111
Sulphuric acid	0.0605
Phosphoric acid	0.0950
Chlorine	0.0053
Total	100.5880

This soil is all that could be desired for the culture of the grape. It contains an abundance of all the mineral substances which enter into the composition of the vine, as shown above by its analysis. While it is *warm*, *light* and *dry*, it contains large quantities of *magnesia* and *vegetable matter*, or *humus*, giving it great capacity for absorbing and retaining a sufficient quantity of moisture, even in the droughts of summer.

This is a fair representative of the soils on the Magnesian Limestone ridges and slopes throughout Central and Southern Missouri. These slopes and ridges occupy millions of acres, now deemed worthless, which are, in fact, by far the most valuable lands in this part of the State for the cultivation of the grape. Especially is this true of those located upon the southern highlands, away from the vapors and sudden changes of our large rivers and their broad valleys.

The Magnesian Limestone series, from which this soil is derived, occupies a large part of the poor portions of the country on the South-western Branch. The Magnesian Limestones, sandstones, porous chert, and the thin beds of reddish, brown marly clays that usually overlie the limestones, all combine to form a soil *light*, *dry*, *warm* and *rich*, in all the elements needed for the grape, as shown by the foregoing analysis. In many places this soil is underlaid with a sufficient quantity of pebbles and fragments of porous chert to constitute a most thorough system of drainage, while in others the fragments of chert are disseminated through the soil in such quantities as to injure it somewhat for ordinary cultivation, but which gives precisely the preparation so highly recommended by Virgil and late authors, and the best cultivators of the grape.

It is true that the native vines do not grow so large and sappy in this as in the deep damp soils of the State; but they are nevertheless strong and healthy, and produce finer clusters of larger and better grapes. This improvement was particularly observed in the *Muscadine*, the *Northern Fox*, and the *Summer Grapes*.

This variety of soil also extends over other portions of the State. It occupies large portions of nearly all the highlands in Southern Missouri, the counties on both sides of the Osage, and over the southern part of Boone, Callaway, Montgomery, and Warren, on the north side of the Missouri, occupying, in all, an area of some 15,000,000 acres. Of these, at least 5,000,000 acres might be selected in the most desirable localities, much of it on the line of the South-western Branch, and devoted to vineyards without encroaching upon the lands most desirable for other departments of agriculture. And, so far as we can judge from the characteristics of soil and climate, and the indications of the native vines, these 5,000,000 acres in the highlands of Southern Missouri, present rare inducements to the vine-dresser—such a combination of favorable circumstances as will not fail to attract the attention of those who would engage in this most pleasant and profitable department of husbandry. And so important will be the results, that every effort should be put forth to hasten the time when these 5,000,000* acres shall be covered with flourishing vineyards; giving profitable employment to 2,000,000 people; yielding more than 1,000,000,000 gallons of wine; and an annual profit, at the lowest estimate, of $500,000,000. And, what is still more important, the pure nourishing juice of the grape would take the place of the vile, maddening compounds used in the names of wine and brandy; drunkenness would give place to sobriety;

* France has about 5,000,000 acres in vineyards. They yield 925,000,000 gallons of wine, besides the 95,000,000 gallons distilled into brandy, and give profitable employment to 2,000,000 of people, mostly women and children.

and our people, nourished by the grape and its pure wines, would become as robust and hardy as they are now daring and indomitable.

Natural Terraces.—The bluffs of the numerous streams in Southern Missouri usually slope back into knobs and ridges, which are frequently surrounded by numerous natural terraces, so regular and uniform that they appear like the work of human hands, as seen in Plate VIII. These terraces are produced by the decomposition of the strata of Magnesian Limestones which form the bluffs. Their height varies from one to six feet, and the width of the top from two to twelve, according to the angle of the slope and the height of the terrace. Their surfaces are nearly level, and are usually covered with a light, warm and rich soil, containing fragments of chert and the decomposing limestone, all wonderfully prepared by nature for the planting of vineyards. These terraces generally surround high, open ridges and knobs, exposed to the free circulation of the dry atmosphere of the region under consideration.

We have as yet observed but one objection to their use for vineyards. In some places the soil does not appear to be sufficiently deep to secure the vine against the effects of droughts. But, as an offset to the want of depth, it always contains large proportions of carbonate of magnesia and humus, which give it great capacity for absorbing and retaining moisture, as these substances possess this capacity to a greater degree than any of the other ingredients of our soils. And, besides, the thinnest soils on these terraces sustain a vigorous growth of prairie grasses, flowers, shrubs and vines, which produce the finest quality of grapes in great profusion.

Caves.—There are numerous spacious caves in all parts of this interesting country. The temperature of those measured ranges between 50° and 60° F. Many of them would make most excellent wine cellars, as their temperature is sufficiently low and uniform to prevent that acidity to which the wines of all temperate latitudes are predisposed.

These facts respecting the *native vines*, the *climate*, the *experience of our vine-growers*, and the *soil*, clearly prove the capacity of Missouri to become the great wine-growing region of our continent. They should encourage those noble spirits

R.B. PRICE DEL. J.M. KERSHAW. ENGRAVER.

BLUFFS OF MAGNESIAN LIMESTONE, ON THE NIANGUA, 3 MILES N.W. OF GUNTER'S SPRING, SHOWING THE NATURAL TERRACES SO COMMON IN THE BLUFFS OF THAT STREAM.

who have so faithfully devoted their labor and their money to promote this important department of husbandry in our midst; for the time is not far distant when the "poor flint ridges" and terraced slopes of Southern Missouri will be as valuable for vineyards as some of them are now for their rich mineral deposits; when the vineyards of Pulaski and La Clede will compete in golden profits with the hemp farms of Lafayette and Platte; and the vine-clad hills of the beautiful Meramec and the Gasconade will vie in wealth with the leaden veins of Potosi and Granby.

It will thus be seen that even the poorest soils and those in the most broken parts of this country will become very valuable for the culture of the grape. And I might add that their value for vineyards will increase in about the same ratio as their fitness for the other departments of husbandry decreases.

ABSTRACTS FROM THE COUNTY REPORTS OF DR. B. F. SHUMARD.

"*Crawford County.*—This county presents great variety of surface, from level or moderately rolling prairie, and 'oak openings,' to rough, rocky hills with abrupt and uneven slopes. The soil varies from rich alluvial bottom land to productive or nearly sterile upland. On the northern side of the dividing ridge, on which is located the Pacific Railroad, the country consists of moderately rolling or level oak openings and prairie, traversed by numerous beautiful prairie valleys, bounded by gentle hills from eighty to one hundred and fifty feet high, the whole presenting a most desirable region for the agriculturalist.

"On the southern side of this ridge the topographical features of the country are quite different. Near the Meramec and its principal affluents, Huzza, Crooked and Dry creeks, the surface is often extremely rough and rocky, and the hills from one to four hundred feet in height. But on the summits of these ridges we frequently find extensive tracts of nearly level, post oak, black oak, and hickory lands, which are capable of being cultivated to good advantage. The alluvial bottoms of all the principal streams are broad, extremely productive, and very heavily timbered.

"*Phelps County.*—In its general features this county is very similar to the preceding. It is generally rolling, and possesses a great deal of fine agricultural land, with here and there districts that are quite broken and illy adapted to cultivation. The western portion is the most broken, particularly in the vicinity of the larger streams. So soon as we leave the valleys of these streams, we encounter rough, rocky hills with abrupt slopes, characterized by poor and sometimes barren soils, extending back for distances varying from a half of a mile to two miles on either side. Then succeed elevated and gently undulating table lands, possessing moderately fertile soils. There is also some rough country bordering the valleys of the Dry Fork of the Meramec and Norman Hollow. The dividing ridge between the Meramec and Bourbeuse presents a succession of beautiful woodlands and prairies, and affords some of the finest farms in the county. On the north side of this ridge we have rolling oak lands, dotted occasionally with patches of prairie. They possess arable soils, particularly where the underlying rock is the 2d Magnesian Limestone, which under proper culture yield abundant and profitable crops. From experiments made in the county by an intelligent farmer, we know that these lands are capable of vast improvement from thorough subsoiling.

"The valleys of Little Piney, Spring and Dry Fork of Meramec and Bourbeuse, have a width varying from a hundred yards to a half of a mile, and their soils are remarkable for their productiveness, throughout nearly their whole extent. The valleys of the smaller streams contain also many very desirable farm sites.

"*Pulaski County* is in general very hilly and broken, but there are extensive districts of rich and productive agricultural lands in the alluvial bottoms of the streams, as well as in the uplands. The hills range from fifty to five hundred feet above the water-courses. If we travel back from the streams, avoiding the valleys of the smaller branches, we usually find at first very rough hills with steep declivities, strewn with a great deal of chert and sandstone, then the surface becomes gently rolling, or expands into level plains, constituting what are known in the country as "*post oak flats*," which are found on the summits of most of the higher ridges, and vary in width from a hundred yards to a couple of miles. For a short time during the spring these plains are occasionally wet, but after they have been once thoroughly broken up by the plough, they are no longer so, but form desirable farming lands. The most extensive "*flats*" lie between the Gasconade, Robideaux, and Big Piney, and east of the latter stream; they also frequently occur on the ridges in the northern part of the county. The valleys of the principal streams are from a few hundred yards to a mile wide, and are remarkable for the fertility of their soils. Indeed, they may be grouped with the very finest soils of our State for the culture of corn, and after being partially exhausted are well adapted to the growth of wheat and other species of small grain. The soils of the smaller valleys are also quite productive, and many of the choicest farms of the county are here located. They do not usually exceed a quarter of a mile in width, but often extend for several miles in length, and are then known as "*Prairie Hollows*." The next in point of fertility are the soils of the uplands, underlaid by the 2d Magnesian Limestone, and the poorest are those overlying the 2d Sandstone and cherty beds of the 3d Magnesian Limestone.

"*La Clede County.*—In its topography this county is very similar to Crawford and Wright, exhibiting great diversity of surface. In the vicinity of the Big Niangua, Gasconade and Osage Fork, the hills range from one hundred and fifty to five hundred feet in height, and are separated from each other by deep and narrow valleys. But after we leave these streams a short distance, the face of the country assumes a less broken aspect, and as we approach the summit level, we find moderately rolling oak lands and broad oak flats, in which may be located many productive and desirable farms. Between the Osage Fork and Gasconade the Pacific Railroad passes over a broad and fertile district of undulating oak openings, interrupted by extensive prairies. The valleys of these streams are from a quarter of a mile to one mile wide. They possess soils whose richness can scarcely be surpassed, and support a heavy growth of the finest kinds of timber. The valleys of the small branches are also highly arable. Those of Goodwin Hollow, Bear, Mill, Cobbs, Prairie, and Brush creeks, afford numerous excellent farms.

"*Wright County.*—The surface is hilly and occasionally rough and broken. The elevation of the hills ranges from fifty to four hundred and fifty feet above the adjacent streams. Most frequently they are neatly rounded in outline, and present gradually ascending slopes. Near the Gasconade and its branches, their sides are often rough and precipitous. The 'Ozark Mountains' (hills would be more proper) traverse the southern tier of townships and constitute the dividing ridge between the waters of the Missouri and White river. The bearing of this ridge is nearly east and west. The ascent from the north is rather moderate, but the southern slope presents usually steep declivities down to the valleys. The soils of the uplands are of course greatly modified by the character of the subjacent strata. Throughout much the largest portion of the county, the soils are of excellent quality, and produce well, while the land is just sufficiently undulating to secure proper drainage. In places where the arenaceous and cherty beds of the Magnesian Limestone series reach the surface, the soil is thin and light, and sometimes entirely unfit for cultivation from the large proportion of chert it contains. The soils of the valleys of every part of the county are remarkable for their richness and fertility."

ABSTRACTS FROM MR. BROADHEAD'S NOTES.

"In Green county, the heavy timber in the bottoms of the Pomme de Terre and the Sac, of James' Fork of White river, of Clear creek, and the Finley, clearly indicate the richness of the alluvial soil in those beautiful valleys. The soil is also good in a part of the country between Stephen's mill and Ray's post office, in sections 17 and 18 of T. 30, R. 24, and in the larger part of Grand prairie, Leaper's prairie and Buck's prairie, in Ts. 26 and 27, Rs. 25 and 26.

"In Lawrence county, in Sections 26, 27, 28 and 36, of T. 26, R. 26, in the valleys of Spring, Crane, Center, and Honey creeks, and in Sarcoxie and Ozark prairies, the soil is excellent.

"In Newton county, the valleys of nearly all the streams are rich and well timbered; that of Indian creek is beautiful and rich, as are also the valleys of Hickory, Shoal and Buffalo creeks.

"In Jasper, the bottoms of Silver creek, and Carver's of Spring river, are rich and well timbered. Round, Dimond and Spring river prairies are rich.

"In La Clede and Camden, the valleys of the Auglaize and the Gasconade, and their tributaries, and many of the slopes descending to those streams, are covered with fine soils.

"In Maries county, the soil is good in the bottom of Spring creek, in the valley of the Maries, and on the Dry Fork of the Bourbeuse, and in Lane's prairie, and the adjacent timbered lands.

"In the following localities the soil is good, though somewhat inferior to that in the places above named:

"Sec. 16, T. 27, R. 21, and Sec. 10 of R. 22, in the same township; the country from Ray's post office to the south-west some five miles; in Sec. 28, T. 27, R. 24, and the valleys at the head of Buck prairie, in Sec. 18, T. 26, R. 24; from the head valley of Spring river, in Sec. 36, T. 26, R. 26, to the head of Crane creek, in Sec. 4, T. 25, R. 25; from Pickerel creek to Sec. 24, T. 29, R. 24; the timbered land between Grand and Leaper's prairies; the hills near Clear creek, in Green county; Secs. 7, 8, 18, 19, 20 and 30 of T. 30, R. 24; Secs. 24 and 25 of T. 30, R. 25, and Sec. 10, T. 29, R. 24; the valleys of the South Fork of Pomme de Terre and the North Fork of the Sac, in Secs. 4, 5 and 6, T. 30, R. 19; Secs. 35 and 36, T. 31, R. 20, and Secs. 1, 2 and 3 of T. 30, R. 20; the ridges between Ozark and Mr. W. C. Smart's, in Green county; the narrow bottoms in T. 24, R. 27; from Sec. 28, T. 26, R. 26 to the south-west, along the Railroad line, some four or five miles; the timber near the head of Little Indian creek, and Sec. 18, T. 24, R. 30; Sec. 8, T. 24, R. 30, and T. 24, Rs. 33 and 34; the southern part of Swa's prairie, in Newton county; the timber from Swa's prairie to Jay's prairie, in the south-east part of T. 25, R. 32; the timber on the edge of Dimond prairie, and on the Railroad line from Hickory creek west to the State line; Spurgeon's prairie, and the county about Duff's lead mines, in Sec. 31, T. 28, R. 32; the bottoms in Sec. 18, T. 27, R. 28, and T. 28, R. 31; from Spring river to Mt. Vernon, and from Mt. Vernon north-west to Adam's branch, and from the Middle to the East Fork of the Turnback; from Pickerel creek to Sec. 9, T. 28, R. 24; the lands on the St. Louis and Springfield road, through the most of Webster and La Clede counties; parts of the valleys of Dry, Spring, Cave-Spring, and

Little Tavern creeks, in Maries county; the hills and valleys near Clifty Dale, and the white-oak lands of Maries county."

MINERALS.

The mineral wealth of the region under consideration is very great, and cannot fail, when fully developed, to command the admiration of the world, and greatly increase the material wealth of our State.

BUILDING MATERIALS

are very abundant in all parts. There is an ample supply of limestones and sandstones and marbles, suitable for all the purposes to which such materials are usually applied. Clays and sands of excellent quality for limes and cements exist in large quantities in nearly all parts of this country. Gravel and pebbles of good quality for roads and streets occur in great abundance.

IRON ORE

of most excellent quality exists in great quantities. The Red and Brown Hematites are the most common; they occur in nearly all the counties, and are found in the Ferruginous Sandstone and the Magnesian Limestones. One of the most valuable localities of iron was observed in the south-western part of Green county. Large masses of fibrous brown hematite cover several acres in the S.E. qr. of the S.E. qr. of Sec. 24, T. 27, R. 24. The bed is more than eight feet thick in a shaft sunk into it. In the S.W. qr., Sec. 19, T. 27, R. 23, we saw another large bed of the same ore. The same excellent ore covers many acres in the N.W. qr. of the same section. It also abounds in Sec. 7 of the same township, and in Secs. 14 and 15, T. 27, R. 24. There are also large beds of this ore to the N. and N.E. of these localities. Some important beds of the common brown hematite occur at Pond Springs, and several other localities in Green county. In Sec. 2, T. 25, R. 25, in Stone county, large quantities of the ore were observed. Beds of less importance were also seen in nearly all the counties examined.

In *Dent county*, in Secs. 2, 3, 10 and 11, of T. 35, R. 4 W., is one of the most valuable and extensive deposits of the specular oxide of iron, near the line of the South-west Branch. The ore is rich and pure, and will yield a very large per cent. of the very best iron. In appearance the ore is intermediate between that of the Iron Mountain and that at the Pilot Knob; but in quality it is not surpassed by either. These beds must become very valuable as the county settles up and the demand for iron, in that part of the State, is greatly increased.

Brown hematite was observed in many localities in *La Clede county*. Mr. Engelmann examined large masses of it near Bear creek, in Sec. 25, T. 36, R. 14. The Meramec Ore Beds, in Phelps county, is a valuable deposit of compact specular ore, which has been wrought since 1829. In Sec. 32, T. 37, R. 8, there is another extensive bed of the same ore. Iron ore is also reported in Sec. 27, T. 36, R. 7; Sec. 11, T. 39, R. 8, and in Sec. 13, T. 37, R. 7.

In *Crawford county* there are many very important localities of iron ore, as shown by the following abstract of Dr. Shumard's report on that county:

"Iron ore of excellent quality has been found at a number of localities in this county, generally associated with the 2d Sandstone and the cherty parts of the 3d Magnesian Limestone. The varieties observed are the brown hematite, specular oxide, and sulphuret. Brown hematite and the specular oxide are found in S.E. of N.E. qr. Sec. 5, T. 37, R. 4 W. It is thickly strewn over the surface, and probably exists in workable quantity.

"Brown hematite occurs at a number of points in T. 36, R. 3 W. In Secs. 15 and 36 it is most abundant, commingled with pseudomorphous crystals of pyrites, chert and crystallized quartz. At Bleeding Hill, according to Mr. Engelmann, there seems to be a rich deposit of specular ore of excellent quality. Two shafts have been sunk here, one of them through thirty-seven feet of red clay and comminuted chert. In this shaft the miners encountered a four-foot bed of *soft, purple iron ore*, greasy to the touch, like the paint ore at the Meramec Iron Works.

"In S.E. of S.E. qr. Sec. 2, T. 38, R. 6, and Sec. 1, T. 38, R. 3, are workable beds of iron ore.

"In N.W. of Sec. 13, T. 37, R. 7, specular ore abounds, together with pseudomorphous crystals of pyrites.

"Specular ore of fine quality abounds in S.E. of S.W. qr. Sec. 32, T. 35, R. 5. Specular oxide is also found in Sec. 4, T. 37, R. 3, and other places in the same township.

"But little mining has been done in this county; still, the surface indications warrant the opinion that the mines are worthy of being more thoroughly tested.

"Iron ore of the best quality abounds at a number of localities in Phelps county. The oldest known, and, perhaps, most valuable deposit in this county, is the Meramec Ore Banks, situated about a half mile from the Meramec, on the west side. This bank was opened as early as 1826, by Messrs. Massey & James, who commenced the erection of a furnace, which was completed in the month of January, 1829, and has been in operation at intervals up to the present time. The ore, which is a rich, compact specular variety, is wrought by Messrs. James, the present proprietors, with considerable profit. It occurs in large rounded or angular masses, and appears to be almost inexhaustible.

"When the masses are broken they exhibit cavities filled with small, extremely beautiful, fibrous crystals of iron, which are highly iridescent, and sometimes perfectly transparent quartz crystals. In some parts of the bank the specular ore is imbedded in a soft, purplish hematite, which is quite soapy to the touch. It forms an excellent and valuable paint, for which purpose large quantities, I am told, are sent annually to the eastern cities. The sandstone in the neighborhood contains masses of iron pyrites.

"In Sec. 32, T. 37, R. 8, there is an extensive deposit of specular ore, very similar in character to the Meramec Bank. It was examined by Mr. Engelmann.

"In N.W. qr. of Sec. 27, T. 36, R. 7, large masses of *specular* and *brown* iron ore abound on the surface. A shaft of fifteen feet has been sunk here, from which a good deal of argillaceous red hematite has been taken.

"In Sec. 11, T. 39, R. 8, small quantities of good hematite occur; and also in Sec. 13, T. 37, R. 7. Beautiful pseudomorphous crystals of iron *pyrites* were found in large masses near Weber's, in the R.R. cut.

"Iron ore is found at many localities in *Pulaski county*. A large deposit of specular iron ore, similar to that used at the Meramec Iron Works in Phelps county, was examined by Mr. Engelmann in Sec. 31, T. 37, R. 12. In N.E. qr. of Sec. 30, T. 36, R. 11, there exists a large deposit of brown hematite. It occurs in the cherty beds of the 2d Sandstone and 3d Magnesian Limestones. Large masses of brown hematite were also observed on the hills of Bee Branch, in T. 37, R. 10. In a cave in Sec. 19, T. 36, R. 8, sulphuret of iron occurs. Sulphuret and brown hematite are also found in Sec. 9, T. 38, R. 13.

"Fragments of brown and specular ore were observed at many points in *La Clede county*, but only in small quantities. Mr. Engelmann observed large masses of brown hematite near Bear creek, in Sec. 25, T. 36, R. 14.

"*Jefferson County.*—In N.E. qr. of Sec. 4, T. 39, R. 4 E., on land belonging to Mr. Prentiss, is a deposit of brown hematite which appears to be of good quality. The ore projects in large masses above the surface of the ground, and the indications are that it exists in workable quantity."

ABSTRACT FROM MR. BROADHEAD'S NOTES.

"Near the line between Webster and Green counties, in Sec. 18, found hematite in fragments, some of them quite large. They occurred in a ravine about the line between the Saccharoidal Sandstone and 2d Magnesian Limestone.

"In Green county, in Secs. 24 and 25, T. 29, R. 24 W., on the summit and sides of a low hill, masses of brown hematite of a good quality were found. The underlying rock, as observed at the nearest locality, was Encrinital limestone.

"Iron ore is abundant in Maries county, occurring both as a hematite and a sulphuret. The sulphuret is found in small quantities in many places. It is most abundant in Secs. 28 and 30, T. 38, R. 9 W., on railroad land. The sulphuret is frequently changed to an oxide on the exposed surface.

"Good specular ore was found abounding in large masses in Sec. 5, T. 39, R. 11 W., associated with argillaceous hematite. This locality is worth exploring.

"Masses of iron ore were found at Vienna. In Sec. 30, T. 41, R. 7, there is a large deposit of argillaceous red hematite, which, I was told, had been used very successfully as a dye stuff. It is easily mined."

But the localities of iron ore are too numerous to be described in detail. The following table gives the important localities known to us:

LOCALITIES OF IRON ORE ON THE SOUTH-WESTERN BRANCH.

The star (*) denotes the localities where the iron is worked.

No. locality.	Name of Mine.	Township.	Range	Section.	County.	Whether on RR. land.	Kind of ore.	Whether worked.	By whom reported.	Remarks.
1		39	6 E	13	Jefferson				B. F. Shumard.	
2	Prentiss.	39	4 E	4	Jefferson				Shumard.	
3		40	3 E	28	Jefferson				Shumard.	
4	Franklin Iron Mining Co.	42	1 E	SE 14	Franklin		Br. hematite & yel. ochre.	*	Dr. Litton.	In mammillary and stalactitic masses.
5	Mrs. Farrar.				Franklin		Hematite.	*	Litton.	
6		42	2 W	NE 17	Franklin		Br. hematite.		Litton.	
7		41	2 E	NE 4	Franklin	R.			Shumard.	
8		41	2 E	NE 6	Franklin				Shumard.	
9		41	1 E	NE 5	Franklin				Shumard.	
10		41	1 E	NE 6	Franklin	R.			Shumard.	
11		35	2 W	8	Crawford				Shumard.	
12		35	3	15	Crawford		Specular.		Engelmann.	
13		35	2	16	Crawford				Shumard.	
14		35	5	SW 32	Crawford				Engelmann.	
15		39	2	NE 28	Crawford	R.			Shumard.	
16		37	3 W	1	Crawford				Shumard.	
17		37	3	4	Crawford		Good spec'lar		Engelmann.	
18		38	6	29	Crawford		Specular.		Shumard.	
19		38	3	1	Crawford		Specular.		Shumard.	
20		36	4 W	NE 26	Crawford		Hematite.		Engelmann.	
21		36	5	Sec. 32	Crawford		Specular.		Shumard.	
22	Collins.	37	4	SE NE 5	Crawford		He.spec.&sul		Shumard.	
23		36	3 W	15 & 36	Crawford		He. & sul'ret.		Shumard.	
24		38	5	10	Crawford	R.			Shumard.	
25		38	5	11	Crawford				Shumard.	
26		38	5	14	Crawford	R.			Shumard.	
27		38	5	15	Crawford				Shumard.	
28	Bleeding Hill,	38	2 W	4	Crawford		Specular.	*	Engelmann.	

LOCALITIES OF IRON ORE ON THE SOUTH-WESTERN BRANCH.—CONTINUED.

No. locality.	Name of Mine.	Township.	Range	Section.	County.	Whether on RR.land.	Kind of Ore.	Whether worked.	By whom reported.	Remarks.
29	Meramec I. W.	37	6 W	1	Phelps		Specular.	*	Shumard.	
30		38	6	SE 29	Phelps		Specular.	*	Shumard.	
31		36	7	NW 27	Phelps		Spec. & hem.	*	Shumard.	
32		36	7	29	Phelps				Shumard.	
33		36	7	23	Phelps				Shumard.	
34		38	8	2	Phelps	R.			Shumard.	
35		37	8	32	Phelps		Specular.		Engelmann.	
36		37	8	12	Phelps	R.			Shumard.	
37		39	8	11	Phelps		Red hematite		Shumard.	
38		37	7	13	Phelps		Hem. & sulp.		Engelmann.	
39		37	12	29	Pulaski				Shumard.	
40		36	11	NE ¼ 30	Pulaski		Br. hematite.		Shumard.	
41		37	12	31	Pulaski		Specular.		Engelmann.	Large deposit.
42		37	10		Pulaski	R. (?)	Hematite.		Shumard.	
43		41	7 W	SW 30	Maries		Red hem. & r.		Broadhead.	Red chalk, used for dyeing.
44		40	8		Maries		[chalk.		Broadhead	Loose fragments.
45	Pilot Knob.	38	8	SE 5	Maries		Sulphuret.		Broadhead.	
46		38	9	28	Maries	R.	Sulphuret.		Broadhead.	Abounds on Railroad land.
47		38	9	30	Maries	R.	Sulphuret.		Broadhead.	Do. do. do.
48		39	9	8	Maries		Hematite.		Broadhead.	
49		39	9	17	Maries		Hematite.		Broadhead.	
50		40	9	2	Maries		Hem. & sulp.		Broadhead.	
51		40	9	17	Maries				Broadhead.	
52		40	9	30	Maries				Broadhead.	
53		38	10	3	Maries		Hematite.		Broadhead.	
54		41	10	32	Maries		Hematite.		Broadhead.	[red clay.
55		39	11	5	Maries		Specular.		Broadhead.	Large masses of good ore, with
56		40	11 W	33	Maries		Hematite.		Broadhead.	
					Maries				Broadhead.	Sulphuret of iron, very common
57		36	14	25	La Clede		Br. hematite.		Shumard.	
58		36	16	5	La Clede				Swallow.	Specular iron ore scattered over the county in small fragm'ts.

59		30	19 W	18	Webster	R.	Oxide.		Broadhead.	
60		30	24	19	Green		Oxide.		Broadhead.	
61		29	24	4	Green		Oxide.		Broadhead.	
62		29	24	24 & 25	Green		Br. hematite.		Broadhead.	In considerable quantities.
63		30	20	26	Green	R.	Hematite.		Swallow.	
64		27	23	W ½ 19	Green		Hematite.		Swallow.	
65		27	23	SW 7	Green		Hematite.		Swallow.	
66		28	23	NW 4	Green		Hematite.		Swallow.	
67		28	23	SW 4	Green	R.	Hematite.		Swallow.	
68		28	23	5	Green		Hematite.		Swallow.	
69		28	23	NW 8	Green	R.	Hematite.		Swallow.	
70		28	23	NW 9	Green		Hematite.		Swallow.	
71		30	23	NE 5	Green		Hematite.		Swallow.	
72		27	24	14 & 15	Green	R.	Hematite.		Swallow.	
73		27	24	23	Green		Hematite.		Swallow.	
74	Smart's.	27	24	E ½ 24	Green	R.	Br. hematite.		Swallow.	
75		28	24	1	Green		Hematite.		Swallow.	
76	Josiah Slogdill	30	25	SE NE 23	Dade		Hematite.		Broadhead.	
77		28	26	SW 5	Lawrence		Hematite.		Swallow.	
78		28	27	NE 1	Lawrence		Hematite.		Swallow.	
79		26 N	23 W	NE qr. 1	Stone		Hematite.		Swallow.	
80		26		SW 6	Stone		Hematite.		Swallow.	
81		25	25	NE NW 2	Stone	R.	Hematite.		Swallow.	
82		29	24	12	Green	R.	Hematite.		Swallow.	
83		44	7 W	29	Osage		Hematite.		Price.	Very extensive beds of hema- [tite.
84		44	7	31	Osage				R. B. Price.	Do. do. do.
85		44	7	32	Osage				Price.	Do. do. do.
86		44	7	33	Osage				Price.	Do. do. do.
87		44	7	34	Osage				Price.	In a cave in small quantities.
88		36	8	19	Pulaski		Sulphate.		Shumard.	Abundant.
89		38	13	9	Pulaski		Sul.&br.hem.		Shumard.	
90		26	28 N	5						

IRON FURNACES.

1	Franklin M. Co	42	1 E	SE ¼ 14	Franklin		Iron.			Steam power.
2	Meramec.	37	6 W	SW ¼ 1	Phelps		Iron.			Water (?) power.

LEAD.

Lead is the great staple of the South-west. Some years ago, I reported this one of the best lead regions of the world. All the subsequent developments have proved the accuracy of that estimate of the mineral treasures of Jasper and Newton counties. Since that time many of the old localities have been more fully developed with great success; and various new mines have been explored with results, in some cases, still more satisfactory.

The mines on Spring river, on Turkey creek, and in Spurgeon's prairie, still promise the most satisfactory returns; while at Granby, on the northern border of Oliver's prairie, the results have been truly wonderful. In the fall of 1854, there was not a cabin on the site where Granby now stands with several thousand inhabitants; and only one shaft had been sunk beneath the soil into the rich mineral veins, which are now penetrated by thousands. Mining at Granby has been most successful, as is evinced by the great number of miners and smelters and merchants, who have there congregated in so short a time, and so far away from the great thoroughfares of travel, and by their contentment and satisfaction with the results of their labors.

Explorations have also been very successful in showing the existence of large quantities of lead in the northern part of Taney county. There are indeed very flattering indications of an abundance of this mineral in all parts of this county which have been examined.

Granby Mines.—So much has been written of these famous mines that it would seem useless to enter into any detailed description of them. The lead is found in somewhat regular leads, or disseminated through the bed of chert, clay, sand and limestone, partially cemented, which overlies the Mountain Limestone. It also occurs in the crevices and cavities of the limestone; and is very frequently disseminated in greater or less quantities through the regular crystalline beds of that rock.

The Sulphuret of Lead or Galena is the most abundant ore of that metal; but the Carbonate is quite common in a few localities, and the Sulphate is sometimes found. Vast quanti-

R. B. Price Del.

Lith. by Schaerff & Bro

GRANBY IN 1857

ties of Galena have been raised from these mines since they were opened, not less than —— pounds. They often discover masses of this ore so large that it is found somewhat difficult to raise them to the surface.

The statistics of one shaft will give an idea of the quantity of ore raised and the profits of mining at this place. Mr. Frazier's shaft, as I am informed, yields 100,000 pounds of galena per month. In one week alone, it yielded 50,000, which at $20 per thousand, would amount to $1,000; deduct $150 for expenses, and the profits of this shaft alone were $850 for that week; they average about $1400 per month.

The miners, collected here from all parts of the world, seemed to be agreed that the Granby Mines are the best they have ever seen. This opinion of the miners, the vast quantity of lead raised in so short a time and in a locality so far from the means of cheap transportation, and the geological features of the country, all unite in proving these mines the most valuable in the world.

The accompanying sketch, Plate XVII., represents a portion of Granby as it appeared in June, 1857.

The *Center Creek Mines*, in Jasper county, are situated on the boder of the prairie extending from Carthage westward to the territory, on a branch leading into Center creek, in Sec. 36, T. 29, R. 33, and Sec. 31, T. 29, R. 32. The following section will give a correct view of the geological features of the locality:

No. 1—10 to 20 feet of chert, limestone and clay, broken and mingled promiscuously, and more or less cemented into a solid mass. The limestone is not so abundant as the clay, and the chert predominates over both.

No. 2—5 feet of regularly stratified bluish crystalline limestone.

No. 3—10 feet same as No. 1, save the limestone is more abundant.

No. 4—(?) of limestone, same as No. 2.

On the east side of the run, the conglomerate of No. 1 is overlaid by eight or ten feet of brown stratified sandstone, which is the same as that at the first locality named. Irregular veins of galena, very variable in thickness, cut through this conglomerate of chert, etc., and through the limestone, in directions approaching an east and west line, and varying from a perpendicular to a horizontal. The galena usually fills the fissure, when it is small, without any vein rock or gang; but when the opening is large, the sheet of mineral runs through the middle, the space on each side being filled with clay and crystals of calcareous spar.

There are several diggings at this place on the west side, on White's and Conovy's land, at some of which numerous shafts have been sunk, from ten to forty feet, and some drifting done. The more important are, Old Diggings, Burnine's, Lunday's, Howard's, Harker's, and Thorp's Diggings. On the eastern side of the run, on Mr. Chenault's land, several openings have been made.

Of the mineral raised at these mines previous to 1854, 270,000 pounds were sold to Harklerode's furnace, 99,074 pounds to Moseley & Co.'s furnace, besides what had been smelted at other places and that still remaining at the mines. I have no statistics showing the amount raised since, but there is no reason to doubt that systematic mining in this locality would be very successful.

Late operations at Reeder, Clinton, and Orchard's Diggings; at Shrewsbury, Orchard & Brother's Diggings; and at Shilling & Co.'s Diggings, have proved very profitable.

Turkey Creek Mines.—No facts have come to light to diminish our confidence in the value of the lead deposits in this locality.

I have nothing new to add respecting *Moseley & Co.'s Mines*, and *Oliver's Prairie Mines*. These works have been discontinued, not however for want of a good prospect of mineral.

Mineral Point Mines.—Late operations at these mines fully sustain the high opinion formed of them in 1854. The following description gives the condition of these mines at that time:

At Mineral Point are several diggings. Harklerode's, so far as I could judge from the miners and the minerals thrown out (for the shaft was full of water), gives great promise of a fine yield. There are two lodes or sheets of mineral lying nearly horizontal in the Carboniferous Limestone. The upper one is made up of galena and carbonate of lead, and chert and clay, mingled together and cemented, and is about one foot thick. The lower is pure galena, from twelve to eighteen inches thick. These lodes have been explored but a short distance.

At Messrs. Fraser & Cavenar's, one shaft, sunk thirty feet, reached a horizontal lode. The fissure in the conglomerate of chert, etc., is four feet, filled with soft clay and galena. Through the middle runs a sheet of galena ten inches thick, and the space on each side is filled with clay and large masses of cog-mineral. This lode had been explored only some fifteen feet. Other diggings at Mineral Point are quite as much esteemed by the miners. Mr. Frazier has a furnace at this point: about sixty shafts have been worked with success and profit.

Duff's Mines give good indications of fine leads of lead.

Taney County Mines are in the Magnesian Limestone Series. All the geological features and the indications of the mines opened, indicate the existence of vast deposits of lead in those parts of the county examined. In S.W. qr. Sec. 11, T. 26, R. 19, several diggings have been commenced, with reasonable indications of success.

Crittenden's Diggings are in N.E. of S.W. qr. Sec. 1, T. 26, R. 19. The prospect is quite good.

Roberd and Hall's Diggings in S. hf. of S.W. qr. Sec. 6, T. 26, R. 18. Good prospect of lead in the upper part of the 3d Magnesian Limestone.

Sheep Diggings is on top of the bluffs of Swan creek, in S.W. of N.E. of Sec. 12, T. 26, R. 19. The mineral at this place was exposed by the sheep. A vein crosses the creek a short distance south-east of these diggings.

Campbell and Han's Diggings is on the 80 acres south of the last locality.

Bray, Buckhart & Co.'s Mines.—Fine prospect of "float mineral" in lower part of 2d Magnesian Limestone.

"*McFadden's Diggings,*" in Sec. 1, T. 26, R. 19; several shafts have been sunk with a good prospect of success.

Peter's Diggings are located ⅓ of a mile south-west of McFadden's.

Hall's Mines are in N.W. of N.W. qr. Sec. 15, T. 26, R. 19. The prospect is good.

Moore's Shaft, in the S.E. qr. Sec. 9, T. 26, R. 19. Some lead has been raised, and all the indications are good.

Goose Diggings are a short distance north-west of the last locality—a fine prospect of float mineral.

Shawnee Diggings, in N.W. of N.E. qr. Sec. 17, T. 26, R. 19, on Shawnee creek; several shafts have been sunk and a few hundred pounds of galena taken out from the middle of the 2d Magnesian Limestone.

The Mines of Price, Bray & Co., in the S.E. qr. of Sec. 9, T. 26, R. 19, are the most important examined in this county. About thirty shafts have been sunk and some drifting done. Lead was found in all of them. The Lead was found in the clay, and in a crevice in the limestone. 20,000 pounds of galena was taken from one shaft only ten feet deep. The whole amount raised, up to June, 1857, was about 170,000 pounds. Since then the yield has been still more encouraging. These mines promise a very large yield of excellent ore. The systematic mining pursued cannot fail to give the most desirable results.

Mr. C. D. Bray, of this company, has a Blast Furnace on Bull creek, in Sec. 6, T. 26, R. 19.

In Webster county, lead has been discovered in several localities. The most important is Harver's mine, on Lost creek, in Sec. 25, T. 27, R. 19.

ABSTRACT FROM DR. SHUMARD'S REPORT.

" *Crawford County*.—The 3d Magnesian Limestone in portions of this county is highly galeniferous. It is frequently characterized by vertical fissures and caverns, some of them of considerable size.

"*Lead Mines, Mineral Hill*, Sec. 32, T. 40, R. 2 W., examined by Mr. Englemann. The hill extends from N.E. of Sec. 32 to the N.E. of Sec. 33. The formation here is the 3d Magnesian Limestone, which is covered with a thick deposit of red clay. The whole side of the hill is marked with shallow diggings, from which immense quantities of ore have been obtained. These mines have been known for more than twenty years—upwards of 1,000,000 pounds of ore has been raised here, and as many as 500 men have been engaged in mining at one time. The mining has all been confined to surface diggings. East of this place, in Sec. 33, is a crevice containing a vein of lead five inches thick, adhering in a vertical sheet to the Magnesian Limestone. In N.E. of N.E. qr. of the same section, lead has been raised.

"*Williams' Mines*, located west of Mineral Hill, in Sec. 32, T. 40, R. 2, were opened in 1851, and up to April, 1854, the amount of ore raised was 202,183 pounds. During the remainder of 1854 there were raised 145,000 pounds. The course of the line of shafts and tunnels bears N.E. and S.W. The mineral was first procured 25 feet below the surface, and the deepest workings were 75 feet below the surface. The mineral is found in fissures of irregular dimensions, varying from two to eight feet in width, and three to four in height. It occurs in veins ranging through red clay, associated with brown hematite, pyrites and ochre. These mines have not been worked to any extent during the last three years.

"Nearly every portion of Secs. 32, 33, and 34, T. 40, R. 2 W., contains more or less lead. In the N.E. qr. of Sec. 1, T. 40, R. 2, there have been raised 10,000 pounds of ore.

"*Wein's Diggings* are located in S.E. of S.E. qr., Sec. 3, T. 38, R. 2 W. During the winters of 1856 and 1857 about 135,000 pounds of ore was raised.

"*Murtrey's Diggings* are situated north of Wein's Diggings, in the same section. A great deal of ore has been obtained here from surface diggings; but no mining has been done lately. On S.E. qr. of S.E. qr., T. 37, R. 2 W., about 200 pounds have been obtained from shallow diggings.

"*Halbert's*, in S.E. Sec., T. 37, R. 4. In 1844 from 3,000 to 4,000 pounds were obtained here from surface diggings.

"*Evans'*, in N.E. of Sec. 3, T. 37, R. 3. In 1856 about 300 pounds were obtained from surface diggings.

"*Ransom's Mines*, (examined by Mr. Englemann,) in Sec. 15, T. 38, R. 2 W. The ore is found in the upper part of the 3d Magnesian Limestone. The mineral is supposed to occur in horizontal sheets, connecting with pockets. About 54,000 pounds of mineral have been obtained here, but no regular mining has been done.

"*Hinch's Mines*, in Sec. 3, T. 38, R. 2. About 500 or 600 pounds of ore have been obtained here. Lead has also been found in many places in this neighborhood.

"*Trask and Garrison's Mines*, near the middle of west line of Sec. 5, T. 36, R. 2 W., have yielded from 10,000 to 15,000 pounds of mineral.

"*Isgrig's Mines*, S.E. of N.E. qr., Sec. 4, T. 39, R. 2 W. A little surface digging has been done here.

"*Sappington's Mines*, N.W. qr., Sec. 1, T. 39, R. 2, were opened in the spring of 1857, since which time they have yielded 55,000 pounds of mineral.

"*Clark's Mines*, in same section, opened in 1853, have yielded 25,000 pounds.

"*Darby's Mines*, also in same section, were opened in 1855. They have been but little worked, and have yielded 7,000 pounds of ore. The last three mines are situated on the same hill, and were examined by Mr. Englemann. The ore occurs in small crevices and pockets, in Magnesian Limestone, and disseminated as float-mineral in red clay, and sometimes adhering to masses of sulphuret and brown oxide of iron.

"*Railroad* or *Coffee Diggings* are located on a spur of the same ridge, in S.W. qr. of Sec. 36, on Pacific Railroad land. Mining was commenced here in 1857, and 5,000 or 6,000 pounds have been raised.

"*Rutledge's Mines*, N.E. qr., Sec. 21, T. 39, R. 2, have been occasionally worked with good success.

"*Red Hills Mines*, N.E. of S.W. qr., Sec. 23, T. 4, R. 2. About 400,000 pounds of lead have been obtained, mostly from the red clay. A few small veins have been discovered in the underlying Magnesian Limestone.

"*Hibler's Diggings*, in N.E. of N.W., Sec. 35, T. 40, R. 2 W. About 10,000 pounds of ore have been obtained. These mines have only been worked irregularly. The ore occurs in float-mineral, in the clay, in crevices and pockets, and in the form of thin sheets penetrating the Magnesian Limestone. Lead in small quantities has also been obtained in Secs. 26 and 27, T. 40, R. 2 W.

"All the above mentioned mines occur in 3d Magnesian Limestone. Lead has also been found at many places in 2d Magnesian Limestone, but only in small quantities.

"*Carbonate of Lead* occurs in small particles at William's Mines and at Mineral Hill.

"*Phelps County.*—In a cave in Sec. 19, T. 36, R. 8, lead occurs in small quantities, in a seam of Barytes, extending from the entrance back for a distance of a hundred yards. In Sec. 35, T. 36, R. 9, a few pounds of mineral have been collected. In Secs. 24 and 32, T. 39, R. 7, a little has been found. At all the localities in this county, the ore occurs in the 3d Magnesian Limestone. In Sec. 8, T. 39, R. 8, in *Maries County*, Mr. Englemann reports the occurrence of lead in 2d Magnesian Limestone. In 1856, about 350 pounds of mineral were obtained here, and on Rocky branch upwards of 2,000 pounds have been raised during one season.

"*Lead* has been found at but few localities in *Pulaski County*, and in small quantities only. A few pieces were picked up in Sec. 6, T. 35, R. 13.

"In *La Clede County*, lead was found in only one locality, in N.W. qr., Sec. 5, T. 36, R. 16. At this place a few shallow excavations have been made in the 3d Magnesian Limestone, and a few pounds of the Sulphuret raised. The ore occurs disseminated through an impure brown iron ore, a few hundred yards distant from what appears to be a dyke of granite.

"In *Wright County* lead ore has been observed, in small quantities, at several localities in the 3d Magnesian Limestone. In S.W. qr., Sec. 11, T. 29, R. 16, Mr. Prock obtained about 150 lbs of ore. It occurs in cavities, associated with calc spar, in the Magnesian Limestone. Float-mineral has been found also in S.E. qr., Sec. 11, and N.E. qr. Sec. 23, in the same township. On Wood's Fork, about a mile and a half east of the Webster county line, lead occurs in sandy textured magnesian limestone rock. It has also been found at several other places in the same strata; it occurs in irregular masses, resting in cavities in the Magnesian Limestone. In Sec. 23, T. 29, R. 14, about 10 pounds of mineral have been found.

"In *Jefferson County*, lead occurs at a number of localities.

"*Gopher Mines* are located on a high ridge of 2d Magnesian Limestone, in S.E. qr. of Sec. 34, T. 41, R. 5 E. They were worked by a company, and have yielded about 120,000 pounds of ore. A great deal of heavy and calc spar was found mingled with the lead at most of the shafts and excavations.

"*Tarpley Mines*, situated in N.E. qr. of Sec. 11, T. 38, R. 4 E., have been fully described by Dr. Litton, in the 2d Report of the Geological Survey. Since his visit, however, during the year 1855, about 123,000 pounds of ore were raised at these mines, and in the spring of 1856 they were worked with eight hands, and yielded 35,000 pounds.

"*Poston and Tyler's Mines*, located in the west half of Sec. 11, T. 38, R. 4 E., yielded during the year 1855 upwards of 90,000 pounds of ore. According to Mr. Daly, every part of Sec. 11 contains more or less lead. At the Daly Diggings, at the head of the Plattin, 60,000 pounds of ore were obtained from a single shaft in 1846–7.

"*Mammoth and Sandy Mines* have yielded large amounts of lead, and a particular account of them is given by Dr. A. Litton, in the 2d Report of the Geological Survey.

"*Howe's Diggings*, situated in Secs. 3 and 4, T. 39, R. 6 E., were discovered in 1840, and have since yielded about 150,000 pounds of ore, most of which was obtained from shallow excavations. It was chiefly smelted at the furnace at Sandy Mines.

"*Yankee Diggings*, situated in Sec. 6, T. 39, R. 6 E. The ore here exists in a fissure whose direction is nearly north and south, and which contains a great deal of heavy and calc spar with some sulphuret of iron. Most of the ore obtained here was from a shaft about seventy feet in depth.

"*McCormick's Diggings*, situated about three-fourths of a mile south of Yankee Diggings, yielded 13,582 pounds of ore during the year 1855. A number of shafts have been sunk here, some of them more than thirty years ago.

"*Lead* has also been found at several other localities in this vicinity; on Mr. Berry's land upwards of 500 pounds of ore have been raised.

"*Garrity & Butcher's Diggings*, in Sec. 12, T. 38, R. 4 E., have not been worked for some years, but are regarded as being excellent mines.

"*Bisch & Daly's Mines*, located in Sec. 7, T. 38, R. 5 E. One shaft has been sunk here to the depth of eighty feet, and a considerable quantity of lead obtained.

"*Bogy's Diggings*, located in E. hf. of S.W. qr. of Sec. 12, T. 38, R. 4, have yielded considerable lead, but workings have been suspended here for several years.

"*Lee's Diggings*, directly south of Mammoth Mines, in Sec. 13, T. 39, R. 3 E., were wrought to some extent about twenty years ago, but nothing has been done here recently.

"*Robinson's Diggings* are located in Sec. 16, T. 39, R. 4 E. In Sec. 10, T. 39, R. 4, on land owned by Gen. Hunt, about 50,000 pounds of ore have been raised.

"*Kelly's Diggings*, in Sec. 5 of same T. and R., have yielded considerable amounts of lead.

"*Frissel's Mines*, situated in N.W. qr. of Sec. 30, T. 40, R. 3 E. According to Mr. Frissel these mines were discovered in 1842, and they have yielded 125,000 pounds of mineral, of which amount 100,000 pounds were raised during the years 1842–3. They have not been wrought to any extent for several years.

"*Nashville Mines*, in N.E. qr. of Sec. 33, T. 40, R. 3 E., have been worked at intervals since 1827, and have yielded up to the present time about 100,000 pounds of ore.

"*Gray's Mines*, located in Sec. 4, T. 39, R. 3 E., were discovered nearly forty years ago, and have been worked at intervals up to the present time. A few shafts have been sunk here, but most of the lead has been obtained from surface diggings. The ore was formerly smelted on the spot by means of a log furnace, the remains of which are still to be seen.

"*Rocky Diggings*, are situated in S.E. qr. of Sec. 5, T. 38, R. 5 E. Some lead has been obtained here, but these mines have not been worked for several years.

"*Miller's Diggings*, situated in the same section as Rocky Diggings, are yielding lead in small quantity."

Mr. Broadhead reports the following from *Maries county:*

"In S.E. qr. of Sec. 20, T. 41, R. 11 W., fragments of lead were found associated with Sulphate of Baryta. Only a few pounds have been found here, and no mining has been done. The rocks here are the lower beds of the 2d Magnesian Limestone.

"In N.W. qr. of Sec. 34, T. 40, R. 11, about 100 pounds of lead have been taken out. It occurs in a vertical opening between the walls of 2d Magnesian Limestone. The walls are about four feet apart and the course of the fissure nearly east and west, a little N.E. and S.W. The lead occurs with iron ore. I was told that there were too thin vertical sheets of iron ore with lead between."

Franklin County.—There has been no opportunity for examining the mines in this county since 1845; but it is known that many new and important localities of lead have been discovered since that time; and there can be no doubt that the long cherished confidence in the richness and extent of its lead deposits, will be fully sustained. When capitalists are prepared to enter upon wise and systematic mining, many of the localities can be worked with great profit. The most reliable information respecting many of these mines may be derived from the following--

ABSTRACT FROM DR. LITTON'S REPORT OF 1854.

"*Golconda Mines*, in Township 43, R. 1 E., Sec. 8. The first digging was done here, probably, in 1830. The mineral was found here at first in the clay, and for the first two years most of the mining was limited to this. In sinking down, a fissure was discovered; the course of which is N. 10° or 15° E. The greatest width of this fissure is three feet. At a point, south from the shafts, and distant 400 yards, the fissure is visible, and has at that point a width of two or three inches. Seven shafts have been sunk on this fissure, the deepest of which is eighteen, and the shallowest twelve feet. The fissure is filled with clay, mineral and calc spar.

"To the east of this fissure, and distant from it but a few feet, is another, with nearly the same course, and having, in some points, a width of four feet. From it, also, has been obtained galena. These fissures are in the second magnesian limestone.

"Under a new lease, parties again commenced working here last May, and with two hands and a working time of not over four months, they report to have obtained 12,000 pounds of mineral. It is to be hoped that the party now engaged will sink their shafts deeper, and properly explore, by drifting and stoping, this deposit, for it presents strong indications of a perpendicular lode.

"*Vallé and Skewes' Mines.*—These are the Cove Mine and Short Lode, on the north, and the Mount Hope Mine, on the south side of the Meramec.

"*The Cove Mine and the Short Lode* are in township 42 N., R. 1 E., Sec. 22, N.W. qr. They are on the side of a high ridge, the height of which is about 200 feet above the level of the valley. This ridge is capped with about fifty feet of sandstone, the lower portion of which is interstratified with magnesian limestone, and beneath which, so far as explored, are heavy-bedded magnesian limestones, intermixed with chert and quartz.

"At the *Cove Mine*, the galena in found in a vertical fissure, whose average width is not over six inches, the course of which is N. 5° E., and with a slight inclination of seven inches to the fathom to the east. This fissure has never yet been found to widen out much over the above average width, but preserves a nearly uniform course and width, so far as explored. This fissure is sometimes filled entirely with galena; at other points, this is accompanied by heavy spar and calc spar; and sometimes these last, with clay, fill it completely.

"The main shaft is about 150 feet deep, at the head of which is a fine exposure of sandstone that extends up to the top of the ridge. South of this, sixty feet, is the bluff shaft, 132 feet deep; and south of this are three other shafts, varying from eighty-eight to fifty feet in depth, and distant from each other from thirty to fifty-eight feet.

"South of main shaft, three levels have been run, connecting with the different shafts; and north, but two have been cut, at a depth from each other of 101 feet into the hill, and extending northwardly to a distance from the main shaft of over 200 feet. Much of the ground has been stopped away from main shaft, south to the Scott shaft, between the first and second levels, and also between the same levels, north of the main shaft. Above the first level, and north of the main shaft, the fissure has been followed up into the sandstone, and has been found well filled with mineral, which, at the time of my visit, was yielding a large quantity of galena. This is not an unimportant part, for though the results of observation in other mining countries would teach us to anticipate a change in the character and productiveness of a vein, in passing from one rock into another of a totally different character, here, at least, is one fact tending to show that the presence of sandstone was not incompatible with the deposition of the galena, and that, perhaps, it is a too hasty generalization to conclude that our lead deposits are only productive within the limits of the magnesian limestones. The mineral is remarkably pure, and among the many specimens examined I found no intermixture with other ores.

"East of Cove Mine 120 yards, and on the same ridge, is another fissure called the Negro Lode. On it have been sunk, on the south side, three or four shafts, the deepest of which is fifty or sixty feet. Its course is nearly N. 10° W. But little work has been done by the present proprietors.

"Two hundred feet east of the Negro Lode is, apparently, another fissure, and running nearly parallel with it. Nothing has been done towards exploring it, excepting to dig some few shallow shafts on the hill side. It is called the Scott Lode.

"*Short Lode.*—This lode is 300 feet east of the Scott, about 280 yards east of the Cove Mine, and on the same hill with them. The lead is found here in fissure, that varies from one inch to two and a half feet in width. Its course is nearly north and south, being nearly parallel with the preceding. The fissure is vertical, and contains, in addition to the ore, the heavy spar, which most frequently accompanies the galena in this fissure. The lead ore is accompanied, sometimes, by sulphuret of zinc. Frequently, cubes of the galena are found encrusted with crystals of the carbonate of lead.

"A considerable amount of systematic mining has been done here. Three shafts—one, ninety feet; one, eighty-five feet; and another seventy-seven feet, have been sunk: levels at three different depths have been run, and the quantity of stoping has been considerable. It has been, and is still, worked with profit.

" On this ridge, which belongs to the third Magnesian Limestone, are three or four fissures passing down perpendicularly, with a course varying but little from a due north and south, and containing galena as far down as they have been explored. They cover a belt of about 300 yards east and west, and though neither on the top nor on the side of the ridge is there scarcely any natural indication of their existence, they are found, under ground, preserving a uniform course to the north, and one has been traced and worked in this direction nearly 300 feet.

" As we pass directly south from the Cove Mine, we travel through the valley of the Meramec, and at a distance of about half a mile we come to a lone, isolated hill, which, from its total disconnection with all others, and its solitary appearance, has been denominated the Lost Hill. This has a height nearly equal to that of the ridge in which the above mines are situated, and in this it is reported that galena has also been found. After leaving the Lost Hill, and travelling nearly due south, we cross the Meramec, and in the bluffs on the south side we again find explorations for lead ore, nearly on a due south line and about two and a half or three miles from Cove Mine.

" *Evans' Lode.**—The first point we reach on this ridge, at which mining has been carried on, is what is known by the name of Evans' Lode. The galena is found here, also, in a vertical fissure, which has a width at some points of two feet. Its course is nearly north and south. It is filled with clay, sulphate of baryta and mineral, and the galena is frequently intermixed with sulphuret and carbonate of zinc. The mining here extends over a distance of 400 feet north and south, and seven shafts, varying from thirty-eight to one hundred and twenty feet, have been sunk, but three of which, however, are connected with levels. The work has not been so systematical nor so regular as at the preceding mines, and this it is reasonable to suppose would be the case, inasmuch as it has not been worked by the proprietor, but has been leased to different parties.

* See Appendix for a further account of this and Casswell Mine.

"By Mr. Evans I am informed that it has yielded about 200,000 pounds of mineral.

"*Mount Hope Mine.*—Farther south and almost joining the above, and not improbably a continuation of it, is the Mount Hope Mine. They are both in the same ridge, the geological character of which is the same as that of the Cove Mine.

"The lead ore is here also found in a vertical fissure, the width of which varies from one inch to two feet. Its course is a little east of north and west of south, with a very slight inclination to the east. Sometimes it is filled entirely with a sheet of galena, and at other points it is found to contain, with lead ore, clay and heavy spar. The ore is sometimes accompanied with the carbonate and the sulphuret of zinc.

"About thirteen shafts have been sunk, varying from twenty feet to one hundred and thirty-three feet in depth. Most of them have been connected by levels, and the mining has extended over a line of nearly 800 feet, north and south.

"Among the debris brought up from the lowest levels at Mount Hope and Cove Mines were some few well-preserved Pleurotomaria and Euomphalus, and one of the most perfect of these last was almost directly in contact with galena.

"The galena found in this mine is accompanied, at some points, with the carbonate and sulphuret of zinc.

"The ore obtained from the Mount Hope, the Short, and the Cove Mines, has been all smelted, since the commencement of operations by the present company in 1849, in a rude reverberatory furnace in the neighborhood of the Cove, and no separate account has been kept of the yield of each mine. The quantity of lead made from 1849 to October of the present year (1854), according to the statement furnished me by Mr. Wm. Skewes, has been 1,947,780 pounds, all the ore having been obtained from the above mines of the company, and the greater part from Mount Hope Mine. The average number of hands employed has been between twenty and twenty-five.

"A blast furnace is now being erected, with which it is intended to smelt the very large quantity of slag that has been accumulating since the company obtained possession of the mines, and which will increase considerably the total amount of lead obtained from these mines during the last five years.

" *Virginia Mine.*—Some two or three miles nearly due south of Mount Hope is the famous Virginia Mine, on the 16th section, in township 41, and range 1 east. This mine was discovered in 1834 or 1835, by Bartlett Brundage, and the fame of it soon attracted to it a number of miners, who obtained the privilege of working lots of twenty-four feet in diameter; and during the first year of its discovery the number engaged in mining is supposed to have been between 200 and 300. The School Commissioners (for it was on the public school land), in order to secure the rent on the mineral obtained, determined to appoint a single smelter, who should be responsible for it; and the number of applicants was so great, that they decided to make the selection by the drawing of lots, when it fell to John Williamson, who, having held it for a short time, sold to C. B. and I. Inge for $7,000. They having retained this office until the autumn of 1835 or '36, disposed of their right for $14,000 to Mr. Clendennin. He held it for about one year, when the mineral having accumulated in such quantities that he could not or did not smelt as fast as it was brought in by the miners, great dissatisfaction was excited, and the miners having rebelled and refused to furnish him the mineral, suit was commenced, the final termination of which was that the lease granted to him was broken. Soon after a number of smelters were appointed by the Trustees of the Public Schools, and at one time there were as many as ten log and three ash furnaces in operation.

"In 1844, the Meramec Company obtained a lease for working the mine and smelting the mineral, with the understanding that they were to buy the miners' rights to the tracts on the lode. They commenced operations actively and energetically, putting up a steam engine and pump, sinking the shafts deeper, running levels, and erecting a furnace; when one of the parties becoming embarrassed in his mercantile business, and another dying, operations were suspended, in 1846, for the want of funds; and since that time little or nothing has been done, while the machinery has been rusting, the buildings decaying, and the shafts and levels been caving in.

"The ore is found more in a vertical fissure, whose course is nearly due north and south, and has been traced by diggings from a short distance north of the Meramec, over a line, extending northwardly into the Bennett tract, of not less than one mile in length. The fissure varies in width from one to fifteen feet; and at one point, at which it is still visible from the top of the shaft, is not less than two feet wide. The rock is covered with a thick, heavy bed of ferruginous clay, the average thickness of which is fifty feet, beneath which is some ten or twelve feet of cherty limestone, and below this is the magnesian limestone. The fissure is filled with clay, heavy spar, (some of which was well crystallized, mostly, however, amorphous, with a light sky-blue color,) and with galena.

"From this section it will be seen that the shafts sunk were very numerous; but, doubtless, before the possession of the mine by the Meramec Company, most of them were sunk without regard to any system or regular mining operations. After the company took possession, the mining was more systematic, and most of their labor was confined to the neighborhood of the engine and north shafts, each of which was sunk to a depth of about 260 feet. Levels were cut from north shaft, both north and south, the latter communicating with Duguid and Prior's shaft. Dr. King, in his report, says that between engine and north shaft there was a vast cavern, extending from the first level connecting these two shafts, almost to the surface of the ground, with an average breadth of nearly five feet, and from fifty to one hundred feet in height, nearly filled with pure galena; and that in the engine shaft, at the depth of 260 feet, the lode was as large and distinct as it generally was throughout the shaft.

"Before the operations of the Meramec Company, the mining was carried on at different points by different parties, acting without regular system, and the one independently of the other. Most of the mineral, I doubt not, was then obtained from comparatively shallow depths. How much of this fissure has been worked out along its course, so far as yet explored, and to the depth of the deepest shafts, I have no sufficient data to enable me to judge; but from the best information I have been enabled to obtain of the levels and the stoping, I should deem it an exaggerated estimate to place it at one-half.

"Of the total amount of mineral obtained here, it is, perhaps, impossible at present to obtain any true and accurate statement. Dr. King, who had an opportunity, about ten years ago, of examining the books of the School Trustees, found the total amount on which rent had been charged and paid, to be 4,610,-158 pounds; but neither he nor any one else supposes that this was all that, up to that time, had been obtained.

"Among all the estimates I have obtained from those who were familiar with the operations at this mine, there is none less than 8,000,000 pounds; some 15,000,000 pounds; but the majority of them place it at 10,000,000 pounds of ore.

"However great may seem the above estimate, I do not doubt, had shafts been sunk systematically, levels been run at suitable and required depths, machinery been erected to keep the mine dry, and the ground been stoped away with any thing like scientific and practical skill, that the Virginia Mine would have been more productive than it has been, and, instead of lying idle, would be still yielding a handsome interest on the investment.

"For many of the above facts, in regard to the Virginia Mine, I am indebted to the Rev. Mr. Clarke and Mr. I. Nash Inge.

"*Darby's Mine*, in Town. 41 N., R. 1 W., Sec. 20, S.E. qr.—This mine was worked some four or five years ago, and, according to all reports, with considerable profit. Operations were suspended on account of the water, but lately a new lease has been obtained by Mr. Giles, who is now engaged in working it.

"This mine is in the spur of a magnesian limestone hill. A shaft has been sunk fifty-two feet deep, and an adit cut for the purpose of drainage. At the bottom of this shaft a level has been run thirty feet, nearly east and west, and near this was found a large cave (denominated by the miners, chimney), extending nearly to the surface of the hill, and which was found filled with clay, tumbling rock, and a considerable quantity of mineral.

"The quantity of water (which is removed by pump, worked by horse power) is so great, that it is necessary to keep the present pump in constant operation, night and day; and, this having been intermitted for several days previous to my visit, I found the shaft filled with water to nearly the adit level.

"Specimens of the mineral seen from this mine were tolerably massive, much of it crystallized in cubes, the sides of many of which were coated with crystals of the carbonate of lead. At the bottom of the shaft were found considerable quantities of the yellow iron pyrites, intermixed with sulphuret of zinc.

"Mr. Giles reports, that during the seven months he has been working, with the assistance of seven hands more than half the time, and during the remainder with that of only four hands, he has obtained 3,000 pounds of mineral. The estimated amounts of mineral, obtained from this mine, anterior to Mr. Giles' lease, varies from 100,000 to 126,000 pounds of mineral.

"*Elliott Mine*, in Town. 41 N., R. 1 W., Sec. 6. This mine lies on the south-western extremity of a ridge, the course of which is a little west of north, and east of south. According to Dr. Shumard, the top of the hill is sandstone, beneath which is the third magnesian limestone.

"The only mineral obtained here has been from the clay, on the side of the hill, one acre of which is almost entirely covered with shallow shafts, the deepest I found open being twenty-one feet. The mineral obtained has been principally from three ranges, the general course of which was N.E. and S.W., running parallel with one another, and distant fifteen to twenty feet from each other. The exposure in the shafts was a reddish ferruginous clay, varying from twelve to twenty feet, below chert, and beneath this the tumbling magnesian limestone. The average depth of the shafts is not over twelve feet, and the deepest ever sunk was forty feet.

"The mineral is a very pure galena, accompanied by neither calc spar nor heavy spar, and exhibits not the least intermixture with either iron or zinc ores. As yet, it has been found only in the clay and chert. Work was commenced here in June, 1853; and since then, with six hands, it is reported that 70,000 pounds of mineral has been obtained.

"Besides the above, there are quite a number of points in Franklin county at which galena has been obtained, and, at some of them, in considerable quantities, but which were not worked during the times of my visits to that county in 1853 and 1854. Most of them were not visited; and I subjoin a list of them, with the amounts of mineral which were reported to me as having been obtained.

"On the school section, in town. 42 N., R. 1 W., in 1827 and

'28, there had been considerable digging. The mineral was found in the clay. The deepest shafts were about fifty feet. The diggings extended over an area of nearly ten acres, but did not extend down into the rock. Mr. A. Chambers, who worked these, obtained and smelted during the above years 40,000 pounds of mineral, and estimates the amount obtained at other times, and hauled to other furnaces, at 25,000 pounds.

"The *Hamilton Mines*, Town. 42 N., R. 1 W., Sec. 31, have not been worked for the last six years. The digging was confined to the clay, and the amount of mineral reported to have been obtained was 100,000 pounds.

"At *Massey's Mine*, Town. 41 N., R. 1 W., Sec. 14, one shaft had been sunk sixty feet, but most of the other shafts were not over twelve feet. Up to October, 1853, Mr. Massey estimated the amount of mineral obtained at from 2,000 to 3,000 pounds. They are much incommoded by water at these diggings.

"*Berthold* and *Generally's Diggings* are near Mitchel's creek, in Sec. 13, Town. 41 N., R. 1 W. They are principally on the side of a hill. The deepest shaft was fifty-four feet, and which was filled with water at the time of my visit, in October, 1853. Mr. Generally gave, as the total amount of mineral obtained here, 100,000 pounds.

REPORTED AMOUNT OF MINERAL OBTAINED.

Silver Holiow	Mines,	Town. 40 N., R. 1 W., Sec. 8,	140,000 lbs.
Thomas'	"	Town. 41 N., R. 1 W., Sec. 32,	100,000 "
Lolla	"	Town. 41 N., R. 2 W., Sec. 15,	50 to 100,000 "
Wheeler	"	Town. 40 N., R. 1 W., Sec. 6 and 7,	50,000 "
Nick Frank's	"	Town. 42 N., R. 1 W., Sec. 8, S.W. qr. of S.E. qr.	
Whitmire	"	Town. 41 N., R. 1 W., Sec. 28,	60,000 "

LEAD FURNACES IN FRANKLIN.

"Formerly, not only in Franklin, but also in other counties in the mining region of Missouri, only the log and ash furnaces were used. These have been gradually replaced everywhere, excepting at one locality in Washington county, by either the Scotch hearth or the reverberatory furnaces. The Scotch hearth requires a blast, hence sometimes called the blast furnace, and this is produced either by water or horse-power, or by steam.

"The old log furnace was simple in its construction, and easily built. After the smelting of one charge, about 5,000 pounds of ore, the furnace was cooled, and after the removal of the ashes, which were rich in lead, it was again charged.

"When, after repeated smeltings with the log furnace, a sufficient quantity of ashes had been accumulated, these were washed to separate the wood from the mineral ashes, when these last were smelted in an ash furnace.

"The slag, from both the reverberatory furnace and Scotch hearth is washed and cleaned, and re-smelted in a slag furnace.

"At present there are but three lead furnaces in operation in Franklin county.

"*Gallaher's Furnace*, T. 41 N., R. 1 W., Sec. 19. It is the Scotch hearth, and the blast is produced by water-power. This furnace has been in operation but two years. All the mineral and slag smelted here came from Franklin county, excepting a lot of 1,900 pounds. Most of the slag came from the Virginia Mine, and Hebbler and Chapman's Furnace.

"Amount of lead made at this furnace, according to the statement furnished me by Mr. Gallaher, was, for—

1853	700 pigs, average of 72 lbs. each	50,400
1854	600 " " " "	36,000

"At the Virginia Mines are two furnaces; only one, however, has, I believe, been in operation since the Meramec Company ceased operations, and this has been under the control of I. Nash Inge.

"*Inge's Furnace.*—According to the statement furnished me by Messrs. Patridge & Co., the agents of Mr. Inge, the following amounts of lead were made at this furnace, from 1849 to 1854:

	PIGS.		POUNDS.
1849, from 20th June	202,	weighed	13,574
1850, " "	3,237,	"	196,744
1851, " "	1,229,	"	80,606
1852, " "	277,	"	18,630
1853, " "	613,	"	39,989
1854, " "	85,	"	5,557

"*Vallé and Skewes' Furnace, at Cove Mine.*—At this furnace has been smelted only the ore obtained from the mines of the Company. I am indebted to Mr. William Skewes for the following statement of lead made at this furnace:

1850,······	5,000 pigs, average weight of each, 61 lbs,	····300,000
1851,········	5,000 " " " " "	····300,000
1852,········	6,000 " " " " "	····360,000
1853,········	9,463 " " " " "	····567,780
1854, to Oct.,	7,000 " " " " "	····420,000

"Mr. Skewes believes that the amount that will be made this year will fully equal that of 1853.

"Statement of the total amount of lead made at the furnaces in Franklin county, from commencement of 1850 to October, 1854:

	1850.	1851.	1852.	1853.	1854.
Gallaher's Furnace,········				50,400	36,000
Inge's Furnace,··········	196,744	80,606	18,630	39,989	5,557
Vallé and Skewes' Furnace,	300,000	300,000	360,000	567,780	420,000
	496,744	380.606	378,630	658,169	461,557"

There are many other localities too numerous to be described even in the brief manner thus far pursued. The accompanying table of mines and localities will give some idea of the numerous localities of lead already known in this district of country; and it is safe to conclude that a small part only of what really exists, has as yet been discovered.

TABLE OF LEAD MINES AND DEPOSITS NEAR THE SOUTH-WESTERN BRANCH.

COMPILED BY MR. G. C. BROADHEAD.

* The star denotes the localities where the lead is worked.

No. of Locality.	Name of Mine.	Town-ship.	Range.	Section.	County.	Whether on R.R. Land.	Kind of Ore.	Whether worked.	By whom reported.	Remarks.
1	Tarpley	38	4 E	N.E. 11	Jefferson,	....	Sulphuret,	*	A. Litton,	Has been much worked.
2	Poston & Tyler	38	4 E	W.½ 11	Jefferson,	....	Sulphuret,	....	Shumard,	
3	Garrity & Butcher	38	4 E	N.½ 12	Jefferson,	....	Sulphuret,	....	Shumard,	
4	Bogy's	38	4	S.W. 12	Jefferson,	....	Sulphuret,	....	Shumard,	
5	Garrity	38	5 E	S.W. 5	Jefferson,	....	Sulphuret,	....	Shumard,	
6	Rocky	38	5	S.E. 5	Jefferson,	....	Sulphuret,	....	Shumard,	
7	Miller	38	5	N.E. 5	Jefferson,	....	Sulphuret,	....	Shumard,	
8	Bisch & Daly	38	5 E	7	Jefferson,	....	Sulphuret,	....	Shumard,	
9	Daly	38	5	11	Jefferson,	....	Sulphuret,	....	Shumard,	
10	Gray's	39	3 E	N.E. 4	Jefferson,	....	Sulphuret,	....	Shumard,	
11	Mammoth	39	3 E	12NW of NW	Jefferson,	....	Sulphuret,	not,	Shumard,	Not been worked since 1852.
12	Lee's	39	3 E	N.W. 13	Jefferson,	....	Sulphuret,	....	Shumard,	
13	Kelly's	39	4 E	S.W. 5	Jefferson,	....	Sulphuret,	....	Shumard,	
14	Skewes & Vallé	39	4 E	16	Jefferson,	....	Sulphuret,	....	Shumard,	
15	Hunt's	39	4	10	Jefferson,	....	Sulphuret,	....	Shumard,	
16	....	39	5 E	33	Jefferson,	....	Sulphuret,	....	Shumard,	
17	....	39	5 E	34	Jefferson,	....	Sulphuret,	....	Shumard,	
18	....	39	6 E	2	Jefferson,	....	Sulphuret,	....	Shumard,	
19	Howe's	39	6 E	N.W. 3	Jefferson,	....	Sulphuret,	....	Shumard,	
20	Howe's	39	6 E	N.E. 4	Jefferson,	....	Sulphuret,	....	Shumard,	
21	....	39	6 E	5	Jefferson,	....	Sulphuret,	....	Shumard,	
22	Yankee	39	6 E	6	Jefferson,	....	Sulphuret,	....	Shumard,	
23	McCormick's	39	6 E	7	Jefferson,	....	Sulphuret,	....	Shumard,	
24	Mead's	39	6 E	10	Jefferson,	....	Sulphuret,	....	Shumard,	
25	....	39	6 E	13	Jefferson,	....	Sulphuret,	....	Shumard,	
26	....	40	3 E	2	Jefferson,	....	Sulphuret,	....	Shumard,	
27	Nashville	40	3 E	28	Jefferson,	....	Sulphuret,	....	Shumard,	

28	......	40	3 E	30	Jefferson...	....	Sulphuret,	....	Shumard,	
29	......	40	3 E	33	Jefferson...	....	Sulphuret,	....	Shumard,	
30	......	40	6 E	19	Jefferson...	....	Sulphuret,	....	Shumard,	
31	Frissel's	40	3 E	30	Jefferson...	....	Sulphuret,	....	Shumard,	
32	......	41	3 E	21	Jefferson...	....	Sulphuret,	....	Shumard,	
33	......	41	4 E	5	Jefferson...	....	Sulphuret,	....	Shumard,	
34	......	41	4	26	Jefferson...	....	Sulphuret,	....	Shumard,	
35	Sandy	41	5	18	Jefferson...	....	{ Sulphuret & Carbon. }	*	Litton,...	Accompanied sometimes with Iron pyrites and Zinc blende.
36	......	41	5	19	Jefferson...	....	Sulphuret,	...	Shumard,	
37	......	41	5	23	Jefferson...	....	Sulphuret,	....	Shumard,	
38	......	41	5	24	Jefferson...	....	Sulphuret,	....	Shumard,	
39	......	41	5	26	Jefferson...	....	Sulphuret,	....	Shumard,	
40	......	41	5	32	Jefferson...	....	Sulphuret,	....	Shumard,	
41	Gopher	41	5	34	Jefferson...	....	Sulphuret,	....	Litton,...	
42	McClenahan's	41	5	35	Jefferson...	....	Sulphuret,	....	Shumard,	
43	Old Mines	..	...	....	Washington	....	....	*	Litton,...	On old mines concessions.
44	Shibboleth	38	3 E	S.W. ¼ 22	Washington	....	....	*	Litton,...	
45	Bellefontaine	38	3	S.E. ¼ 9	Washington	....	....	*	Litton,...	
46	Cannon	38	3	....	Washington	....	....	*	Litton,...	Some Iron pyrites combined.
47	Scott & Kee's-	38	2 E	22	Washington	....	....	*	Litton,...	
48	Burt's	37	2 E	....	Washington	....	....	*	Litton,...	
49	Price & Willoughby	37	2 E	15	Washington	....	....	*	Litton,...	
50	New Diggings	37	3	....	Washington	....	....	*	Litton,...	S.E. of Potosi.
51	Lupton	38	2	17	Washington	....	....	*	Litton,...	
52	......	38	2	16	Washington	....	....	*	Litton,...	
53	Casey & Clancy	38	2	7	Washington	....	Sulphuret,	*	Litton,...	Six shafts were worked.
54	Cook's	38	2	....	Washington	....	....	*	Litton,...	
55	Brock's	38	1	N.W. ¼ 4	Washington	....	....	*	Litton,...	
56	Shore's	38	2	S.W. ¼ 18	Washington	....	....	*	Litton,...	
57	Fourché à Renault ..	38	2	7 & 8	Washington	....	....	*	Litton,....	Five different diggings under [this name.
58	Prairie	38	3	4	Washington	....	....	*	Litton,....	
59	Wet & Elliott	39	2	26	Washington	....	....	*	Litton,....	Seven shafts.
60	Rocky	36	3	....	Washington	....	....	....	Litton,....	Many diggings. Not many at [present worked.
61	Faquaher	36	3	....	Washington	....	....	*	Litton,....	

Table of Lead Mines and Deposits—CONTINUED.

No. of Locality.	Name of Mine.	Town-ship.	Range.	Section.	County.	Whether on R. R. Land.	Kind of Ore.	Whether worked.	By whom reported.	Remarks.
62	Pigeon-Roost and Strawberry ...	36	1 W	1	Washington	....	....	....	Litton, ...	Fourch à Courtois Mines.
63	Ismael	36	1 E	E. ½ S.E. 8	Washington	....	....	....	Litton, ...	" These mines have been worked more or less for the last forty years.
64	Madden Hill	36	1 W	S.E. 2	Washington	....	....	....	Litton, ...	"
65	Flint Hill	36	1 W	S.W. 12	Washington	....	....	....	Litton, ...	"
66	Bit	36	1 W	S.W. ¼ 12	Washington	....	....	....	Litton, ...	"
67	Bluff	36	1 W	E. ½ N.W. 14	Washington	....	....	....	Litton, ...	"
68	Water Hill	33	1 W	E. ½ N.W. 15	Washington	....	....	....	Litton, ...	"
69	English	36	1 W	N.E. ¼ 2	Washington	....	....	....	Litton, ...	"
70	Cochran	36	1 W	W. ½ N.E. 12	Washington	....	....	....	Litton, ...	"
71	Picayune	37	1 W	W. ½ S.E. 34	Washington	....	....	....	Litton, ...	"
72	Polecat	36	1 W	W. ½ S.W. 14	Washington	....	....	....	Litton, ...	"
73	Tarkey	36	1 E	15	Washington	....	....	....	Litton, ...	"
74	Gopher	36	1 E	E. ½ S.W. 22	Washington	....	....	....	Litton, ...	"
75	Peru	36	1 E	W. ½ N.E. 19	Washington	....	....	....	Litton, ...	"
76	Clemins	36	1 E	W. ½ N.W. 17	Washington	....	....	....	Litton, ...	"
77	Coffee-Pot	36	1 W	W. ½ N.W. 34	Washington	....	....	....	Litton, ...	"
78	Trash	36	1 E	E. ½ S.W. 6	Washington	....	....	*	Litton, ...	"
79	Sweassey	36	1 E	E. ½ S.E. 18	Washington	....	....	*	Litton, ...	"
80	Montgomery	36	1 E	S.W. ¼ 7	Washington	....	....	*	Litton, ...	"
81	Hypocrite	86	1 E	E. ½ N.W. 22	Washington	....	....	....	Litton, ...	
82	Graveyard	36	1 E	E. ½ N.E. 8	Washington	....	....	....	Litton, ...	
83	Turkey-Hill	36	1 E	W. ½ N.E. 7	Washington	....	....	....	Litton, ...	
84	Richwoods	40	2	....	Washington	....	....	....	Litton, ...	More or less mining done for the last forty years.
85	" La Beaume	40	2	32	Washington	....	....	....	Litton, ...	
86	" French Diggings	40	2	21 & 28	Washington	....	....	....	Litton, ...	
87	Mount Hope	41	1	1, 3 & 4	Franklin ...	....	Sulphuret.	*	Litton, ...	
88	Virginia	41	1	16	Franklin ...	....	....	*	Litton, ...	But little worked during late [years.
89	....	41	1	34	Franklin ..	R	....	....	Litton, ...	
90	....	41	2	34	Franklin ...	R ?	....	....	Litton, ...	R. R. land in this section.

91	{ Cove	42	1	21	Franklin	....	Sulphuret,	*	Litton	
92	} Short Lode	42	1	22	Franklin	R ?	Sulphuret,	*	Litton	Railroad land in this section.
93	Casswell & Evans'	42	1	34	Franklin	....	....	*	{ Swallow & Litton }	Railroad land in this section.
94	Golconda	43	1 E	8	Franklin	R ?	....	*	Litton	7 shafts. Railroad land in this [section.
95	....	40	1 W	5	Franklin	....	....	....	Litton	
96	Wheeler	40	1	6	Franklin	R ?	....	*	Litton	Railroad land in this section.
97	"	40	1	7	Franklin	....	....	*	Litton	
98	....	40	1	8	Franklin	R ?	....	....	Litton	Railroad land in this section.
99	Elliot's	41	1 W	6	Franklin	....	Sulphuret,	*	Litton	
100	Massey	41	1	14	Franklin	R ?	Sulphate,	*	Litton	Railroad land in this section.
101	Generally's	41	1	13	Franklin	....	....	*	Litton	
102	Darby	41	1	20 S.E. ¼	Franklin	....	{ Suphur. & Carb. }	*	Litton	N. ½ section Railroad land.
103	Whitmire	41	1	28	Franklin	....	....	*	Litton	Railroad land in this section.
104	....	41	1	29	Franklin	....	....	....	Litton	
105	....	41	1	30	Franklin	....	....	....	Litton	Railroad land in this section.
106	Thomas	41	1	32	Franklin	....	....	....	Litton	Railroad land in this section.
107	....	41	1	33	Franklin	....	....	....	Litton	
108	Franks	42	1	8	Franklin	....	....	....	Litton	
109	....	42	1 W	16	Franklin	....	....	*	Litton	
110	Hamilton	42	1 W	31	Franklin	....	....	*	Litton	
111	....	44	1 W	24	Franklin	....	....	not	Litton	
112	Lolla	41	2 W	15	Franklin	....	....	....	Litton	
113	Silver Hollow	40	1 W	8	Franklin	....	....	*	Litton	Railroad land in this section.
114	....	36	2 W	7	Crawford	....	....	*	Shumard	
115	Matthews	37	2 W	14	Crawford	....	....	....	Shumard	
116	McKane's	37	2	S.E. 2	Crawford	....	....	....	Engelmann	
117	....	37	2	17	Crawford	....	....	....	Shumard	
118	....	37	2	32	Crawford	....	....	....	Shumard	
119	Hinch	38	2 W	3	Crawford	....	....	*	Engelmann	Railroad land in this section.
120	Ransom's	38	2	15	Crawford	....	....	*	Engelmann	
121	Williams	38	2	30	Crawford	R	{ Sulphur. & Carb. }	*	Shumard	
122	Darby's	39	2 W	1	Crawford	....	....	*	Shumard	
123	....	39	2	6	Crawford	....	....	....	Engelmann	

Table of Lead Mines and Deposits—CONTINUED.

No. of Locality.	Name of Mine.	Town-ship.	Range.	Section.	County.	Whether on R. R. Land.	Kind of Ore.	Whether worked.	By whom reported.	Remarks.
124	....	38	2 W	33	Crawford ..	....	....	....	Shumard ..	
125	Clark's	39	2 W	1	Crawford ..	....	....	*	Engelmann	
126	Sappington's	39	2 W	N.W. 1	Crawford ..	....	....	*	Engelmann	
127	Isgrig's	39	2	S.E. of N.E.4	Crawford ..	....	....	*	Engelmann	
128	....	39	2	2	Crawford ..	....	....	....	Shumard ..	Railroad land in this section.
129	Red Hill	40	2 W	23	Crawford ..	....	....	*	Engelmann	
130	Trask & Garrisson .	36	2 W	5	Crawford ..	....	....	*	Shumard ..	
131	....	36	2	27	Crawford ..	....	....	....	Shumard ..	
132	....	40	2 W	26 & 27	Crawford ..	....	....	....	Engelmann	
133	....	36	2	31	Crawford ..	....	....	....	Shumard ..	
134	Mineral	40	2	32	Crawford ..	....	Sulphur.	*	Engelmann	Railroad land in this section.
135	Hill	40	2	33	Crawford ..	....	& Carb.	*	Engelmann	
136	Hibler's	40	2 W	35	Crawford ..	....	Sulphuret.	....	Shumard ..	
137	Railroad	40	2	36	Crawford ..	R.	Sulphuret.	....	Engelmann	
138	....	36	3 W	16	Crawford ..	....	Sulphuret.	....	Shumard ..	
139	....	36	3	33	Crawford ..	....	Sulphuret.	....	Shumard ..	
140	....	36	3	34	Crawford ..	....	Sulphuret.	....	Shumard ..	Railroad land in this section.
141	....	37	3 W	13	Crawford ..	....	Sulphuret.	....	Shumard ..	
142	....	37	3	14	Crawford .	....	Sulphuret.	....	Shumard ..	
143	Rutledge	39	2	N.E. 1	Crawford ..	....	Sulphuret.	*	Engelmann	
144	....	39	3 W	1	Crawford ..	....	Sulphuret.	....	Shumard ..	
145	Halbert's	37	4 W	1	Crawford ..	....	Sulphuret.	....	Engelmann	
146	Evans'	37	3	3	Crawford ..	....	Sulphuret.	....	Engelmann	
147	....	36	8 W	19	Phelps	....	Sulphuret	....	Shumard ..	
148	....	39	8 W	11	Phelps	....	Sulphuret.	....	Shumard ..	
149	....	39	8	14	Phelps	....	Sulphuret.	....	Engelmann	
150	....	36	9 W	35	Phelps	....	Sulphuret.	....	Shumard ..	
151	....	39	7 W	24	Phelps	....	Sulphuret.	....	Engelmann	
152	....	38	7	32	Phelps	....	Sulphuret.	....	Shumard ..	
153	....	41	11 W	20	Maries	....	Sulphuret.	....	Broadhead.	Found in small quantities.

154	....	39	8 W	8	Maries.....	....	Sulphuret..	....	Engelmann	In small quantities.
155	....	40	11 W	34	Maries.....	....	Sulphuret..	not now,	Broadhead.	In small quantities.
156	....	36	16 W	5	La Clede ..	....	....	....	Shumard..	
157	....	27	19 W	S.W. 25	Webster ...	....	....	....	Swallow ..	
158	Robards, Hall & Co.	26	18 W	S.W. 6	Taney.....	....	Sulphuret..	....	Swallow ..	
159	....	26	18	N.W. 7	Taney.....	....	Sulphuret..	....	Swallow ..	
160	Crittenden	26	19 W	S.W. 1	Taney.....	....	Sulphuret..	....	Swallow ..	
161	Sheep	26	19	N.E. 12	Taney.....	....	Sulphuret..	....	Swallow ..	
162	....	26	19 W	S.W. N.E. 12	Taney.....	....	Sulphuret..	*	Swallow ..	Vein across creek.
163	....	26	19	S.E. S.E. 12	Taney.....	....	Sulphuret..	....	Swallow ..	
164	Bray & Co.	26	19	S.E. S.E. 13	Taney.....	....	Sulphuret..	*	Swallow ..	
165	....	26	19	S.W. 11	Taney.....	....	Sulphuret..	....	Swallow ..	
166	Goose	26	19	S.E. 9	Taney.....	....	Sulphuret..	*	Swallow ..	
167	Hall's..	26	19	{ N.W. 15 & N.E. 16 }	Taney.....	....	Sulphuret..	*	Swallow ..	
168	Shawnee	26	19	N.E. 17	Taney.....	....	{ Sulphur. & Carb. }	*	Swallow ..	
169	Vaughan's	26	21	N. 2	Taney.....	....	Sulphuret..	....	Broadhead.	
170	Phelps	29	21	S. 36	Green.....	....	Sulphuret..	not now,	Broadhead.	Abandoned.
171	Vaughan's	27	21	S.E. S.E. 35	Green	....	Sulphuret..	....	Broadhead.	Abandoned.
172	T. H. Williams.....	29	25	N.E. 2	Lawrence ..	....	Sulphuret..	....	Broadhead.	
173	Richey's	26	25	S.E. 27	Lawrence ..	....	Sulphuret..	*	Price	
174	Upshur	25	25	N.E. 19	Barry	...	Sulphuret..	....	Swallow ..	
175	Spurgeon's	26	32	S.E. 31	Newton	....	Sulphuret..	*	Price	Many shafts.
176	Ailsworth's	26	32	N.W. 32	Newton	....	Sulphuret..	....	Price	
177	....	26	32	31	Newton	....	Sulphuret..	....	Price	
178	Loveley's	25	33	1	Newton	....	Sulphuret..	....	Price	
179	"	26	33	36	Newton	R	Sulphuret..	....	Price	
180	Ryan's	26	32	35	Newton	....	Sulphuret..	not now,	Broadhead.	Formerly Moseley's mines.
181	....	25	30	S.E. S.W. 1	Newton	....	Sulphuret..	....	Swallow ..	
182	Dorris & Cavender .	25	30	S.W. 1	Newton	....	Sulphuret..	*	Swallow ..	
183	Foster	25	30	S.E. 2	Newton	R	Sulphuret..	*	Swallow ..	
184	Granby	25	30	6	Newton	R	{ Sulphur. & Carb. }	*	Swallow ..	Several hundred shafts.
185	Booth, Ryan & Co. .	25	31	S.E. N.E. 22	Newton	R	Sulphuret..	*	Swallow ..	

Lead has been found in Pulaski county, at only a few localities and in very small quantities.—*Shumard.*

Table of Lead Mines and Deposits—CONTINUED.

No. of Locality	Name of Mine.	Township.	Range.	Section.	County.	Whether on R. R. Land.	Kind of Ore.	Whether worked.	By whom reported.	Remarks.
186	Oliver's Prairie	25	30	....	Newton	....	Sulphuret	....	Swallow	
187	Richardson & Brock	..	...	....	Newton	....	Sulphuret	....	Swallow	
188	Davis & Cole	..	...	....	Newton	....	....	....	Swallow	
189	Richardson & Foster	..	...	....	Newton	....	....	....	Swallow	
190	Vickory & Johnson	..	...	....	Newton	....	....	....	Swallow	
191	Strickland	..	...	....	Newton	....	....	....	Swallow	
192	Baxter	25	33 W	4	Newton	R	Sulphuret	....	Swallow	
193	....	25	33	21 & 22	Newton	R	Sulphuret	....	Swallow	
194	....	26	33	9	Newton	....	Sulphuret	....	Swallow	
195	Orchard's	27	33	N.E. 21 ?	Newton	....	Sulphuret	....	Swallow	
196	....	28	31	8	Jasper	....	Sulphuret	....	Swallow	
197	Center Creek	29	33	36	Jasper	....	Sulphuret	....	Swallow	
198	"	29	32	31	Jasper	....	Sulphuret	....	Swallow	
199	Old Diggings	..	...	....	Jasper	....	Sulphuret	....	Swallow	} Center Creek Mines.
200	Burnine's	..	...	....	Jasper	....	Sulphuret	....	Swallow	
201	Sunday's	..	...	....	Jasper	....	Sulphuret	....	Swallow	
202	Howard's	..	...	....	Jasper	....	Sulphuret	....	Swallow	
203	Harker's	..	...	....	Jasper	....	Sulphuret	....	Swallow	
204	Thorp's	..	...	....	Jasper	....	Sulphuret	....	Swallow	
205	Mineral Point, or Shake Rag, incl. several diggings,	28	33	33	Jasper	....	Sulphuret	*	Price	50 or 60 shafts. } Turkey Cr. Mines.
206	Duff's	28	32	S.E. ¼ 36	Jasper	....	Sulphuret	....	Price	4,500 pounds have been raised.
207	Orchard's	27	33	16 ?	Jasper	....	Sulphuret	....	Price	
208	Cox's	27	33	....	Jasper	....	Sulphuret	....	Price	
209	Rader, Clinton & Orchard	..	...	....	Jasper	....	Sulphuret	....	....	500 to 1,000 pounds per hand a day.
210	Shrewsbury, Orchard & Bros	..	...	....	Jasper	....	Sulphuret	....	....	
211	L. Shilling & Co.	..	...	....	....	....	....	....	....	600 to 800 pounds per hand a day.

212	....	29	12 W	1	Wright....	....	Sulphuret..	*	Shumard..
213	....	29	12	2	Wright....	....	Sulphuret..	*	Shumard..
214	....	29	12	10 & 11	Wright....	....	Sulphuret..	*	Shumard..
215	....	29	12	23	Wright....	....	Sulphuret..	*	Shumard..
216	....	29	12	24	Wright....	....	Sulphuret..	*	Shumard..

CATALOGUE OF LEAD FURNACES NEAR THE SOUTH-WESTERN BRANCH.

No. of Furnace.	Name of Furnace.	Township.	Range.	Section.	County.	Ore.	By whom reported.	Remarks.
1	Higginbotham	39	3 E	27	Washington	Lead	Litton......	Log and Ash.
2	T. & W. Murphey's......	39	3 E	S.W. ¼ 28	Washington	Lead	Litton......	Scotch hearth; blast by water power.
3	Long's, Old Mines	......	.....	....	Washington	Lead	Litton......	Scotch hearth; horse power.
4	White's, Old Mines	......	.....	....	Washington	Lead	Litton......	Double Scotch hearth; water power.
5	McIlvaine's	37	2 E	15	Washington	Lead	Litton......	Scotch hearth; water power.
6	Deane's	......	.....	....	Washington	Lead	Litton......	Scotch hearth; water power.
7	Kennett's, at Shibboleth ...	......	.....	....	Washington	Lead	Litton......	
8	Boase's, on Miller...... ...	......	.....	....	Washington	Lead	Litton......	Scotch hearth and slag.
9	Hopewell......	......	.....	....	Washington	Lead	Litton......	Scotch hearth and slag, used for thirty years.
10	Walton's	......	.....	....	Washington	Lead	Litton......	Mostly made from slag, used for fourteen years.
11	Manning's—Webster	......	.....	....	Washington	Lead	Litton......	Scotch hearth; log and ash, from 1831 to 1836.
12	Creswell's	......	.....	....	Washington	Lead	Litton......	Scotch hearth; water power; used since 1835.
13	Casey & Clancey......	......	.....	....	Washington	Lead	Litton......	Scotch hearth; water power,
14	Richwood's......	......	.....	....	Washington	Lead	Litton......	
15	Virginia	41	1 E	16	Franklin...	Lead	Shumard ...	Not now worked.
16	Gallagher's	41	1 W	19	Franklin...	Lead	Litton......	Scotch hearth; water power.
17	Inge's......	......	.....	....	Franklin...	Lead	Litton......	Scotch hearth; water power.
18	Vallé & Skewes—Cove Mine	......	.....	....	Franklin...	Lead	Litton......	Scotch hearth; water power.
19	Sandy	41	5 E	N.W. 18	Jefferson...	Lead	Litton......	
20	Mammoth	......	.....	....	Jefferson...	Lead	Litton......	Scotch hearth; blast—steam.
21	Vallé	......	.....	....	Jefferson...	Lead	Litton......	Scotch hearth and slag; blast—steam.
22	Bray's, Five Points	26	19 W	N.W. 6	Taney.....	Lead	Swallow	Scotch hearth; water power.
23	Blow & Kennett	25	30 W	6	Newton	Lead	Swallow	Steam—6 eyes. At Granby, on Railroad land.

Catalogue of Lead Furnaces—CONTINUED.

No. of Furnace.	Name of Furnace.	Town-ship.	Range.	Section.	County.	Ore.	By whom reported.	Remarks.
24	Dale & Fitzgerald	25	30	6	Newton	Lead	Swallow	Steam, 3 eyes. } At Granby, on Railroad land.
25	W. E. Long	25	30	6	Newton	Lead	Swallow	Steam, 2 eyes. } At Granby, on Railroad land.
26	Johnson & Gregory	25	30	6	Newton	Lead	Swallow	Horse power, 1 eye. } At Granby, on Railroad land.
27	Jno. Plummer	26	30	....	Newton	Lead	Swallow	Water power, 1 eye. Two miles north of Granby, on Shoal creek.
28	Booth, Ryan & Co.	25	31	S.W. 23	Newton	Lead	Swallow	1 eye. Hickory creek, 5 miles south of Granby.
29	C. Legendre & Co.	..	.....	....	Newton	Lead	Swallow	3 eyes. On Hickory Cr., 4 ms. south of Granby.
30	Moseley	26	32	22	Newton	Lead	Swallow	1 eye. 9 miles west of Granby.
31	Harklerode	28	33	N.E. S.E. 10	Jasper	Lead	Swallow	Blast—water power.
32	Hibler's	40	2 W	....	Crawford	Lead	Shumard	
33	Spurgeon's	26	32	N.E. 31	Newton	Lead	Price	Log furnace.
34	Mineral Point	28	33	33	Jasper	Lead	Price	

COPPER.

We observed several localities in the South-west, in which were found small quantities of both the *sulphuret* and the *carbonate* of *copper;* but none of them give evidence sufficient to justify the opinion that the deposits are very extensive.

In *Taney county* small quantities of copper ore were observed at McFadden's Mines and at the Goose Diggings. In Lawrence county, at several localities on the Turnback, this ore was found near the junction of the Encrinital and Chemung rocks.

Mr. Broadhead examined several localities in Dade and Green. The following is an abstract of his report:

REPORT OF LOCALITIES EXAMINED BY G. C. BROADHEAD.

"In *Dade county*, at Josiah Stogdill's, on S.W. of N.W. qr. Sec. 2, T. 30, R. 25, copper ore occurs in small crystals of sulphuret and particles of green carbonate, profusely disseminated through a very coarse and somewhat friable crystalline limestone, belonging to the lower Encrinital beds. Along a branch running near, copper has been found in several places. On S.E. of N.E. qr. Sec. 23, T. 30, R. 25, sulphuret of copper is found, associated with brown hematite.

"*Green county.*—On Sec. 19, N.W. of S.W. qr. T. 30, R. 24, very small traces of copper were found associated with calc spar, and traversing the lower silicious beds ('Turnback rocks') of the Encrinital limestone in about an E. and W. direction.

"At *William Haralson's*, on W. hf. of Sec. 10, T. 29, R. 24 W., a pit has been sunk fourteen feet deep through the lower beds of the Encrinital limestone. The ore found here is the sulphuret and green carbonate, in a gangue of coarse opaque buff-colored calc spar, adhering to large crystals of white subtransparent calc spar, the copper ore more often occupying the line between the two varieties of spar. Some mining has been done here, but no profitable results have as yet been derived. This shaft was sunk in the edge of a valley leading into the valley of Sac river, and about three-fourths of a mile from that stream. Fragments of copper ore have been found at several places along this valley. In Sec. 2, T. 40, R. 9 W., in *Maries county*, some explorations for copper have been made; but proving unprofitable, the mining was abandoned. A very little copper ore was found here, associated with iron pyrites and hematite and dog-tooth spar. The mining extended through red clay, into the softer beds of the underlying 3d Magnesian limestone."

ABSTRACT FROM DR. B. F. SHUMARD'S REPORT.

"The copper mines of Crawford county have not been worked for some years. Dr. H. King examined them at the time they were being worked, from whose report we largely avail ourselves.

"*Hinch's Copper Mines*, on the side of a high hill, near the center of Sec. 4, T. 38, R. 2 W. This mine was discovered in 1849, and several thousand pounds of ore have been raised here. According to Dr. King, the ore, near the surface, is a carbonate and oxide, but deeper it assumes the character of a sulphuret of excellent quality. Dr. King states that 800 lbs. of ore produced 273 lbs. of good pig copper. The holes or shafts have been sunk chiefly in loose, red clay and comminuted chert, but the walls of some of them are in the Magnesian limestone. The copper ore was found with brown hematite in small fragments disseminated through the clay and filling fissures in sandstone. Small scales of native copper were found occasionally with the carbonate and oxide.

"Mr. Engelmann states that very little has been done here toward investigating the real character of this mine, owing to the very irregular manner in which the work has been carried on.

"*Rives' Copper Mine*, in N.E. qr. Sec. 13, T. 39, R. 3 W. The formation here is the cherty portion of the 3d Magnesian Limestone and 2d Sandstone. This mine was worked to some extent in 1849, and many pits were sunk through the superficial deposits. According to Dr. King's report, some twelve or fifteen holes were sunk, and more or less copper in some condition was found in nearly all of them. On the west side of the hill, at a depth of about twenty feet, a mass of ore was struck several feet in thickness, or which was penetrated to this extent without passing through it. Dr. King further states that a 'large pile, probably some hundred thousand pounds of this ore, was brought to the surface, where it has since been left exposed to the rains and atmospheric influences.' It is chiefly an oxide and sulphuret of iron and copper, but not very rich in the latter mineral.

"In most of the other shafts sunk at this mine, the ore was found in the state of green carbonate; but this was generally in a deposit of fragmentary chert.

"Dr. King arrives at the conclusion that this would be an extremely valuable copper mine if properly worked.

"*Copper Hill.*—No work has been done here since Dr. Litton examined it.

"*Bleeding Hill*, in S.W. of N.W. qr. Sec. 4, T. 38, R. 2 W., was examined by Mr. Engelmann. A few shallow shafts have been sunk here, chiefly through red clay and chert.

"The ore is found in small fissures in 2d Sandstone, and consists of green and blue carbonate, sulphuret, and some scales of virgin copper, commingled with a great deal of earthy brown hematite. No systematic mining has been done here, but much useless labor has been spent.

"In Sec. 22, T. 40, R. 2 W., some excavations have been made, but only small fragments of blue and green carbonate have been found. A few pieces have also been found on Huzza and Crooked creeks.

"*Copper*, in small quantities, was found in Phelps county, in the lead cave on Piney, above mentioned.

"In Sec. 23, T. 28, R. 16, in Wright county, Mr. Hough found small particles of green carbonate of copper disseminated through the calcareous spar which here abounds in the Magnesian limestone."

EXTRACT FROM DR. LITTON'S REPORT.

"*Stanton Copper Mine*, T. 40 N., R. 2 W., Sec. 2, where mining was commenced by the present company, in 1851, and has been continued to the present date, without interruption.

"This mine is in the spur of a ridge, the course of which is about N., 70° E., terminating, at its eastern extremity, in a valley. In most places, this ridge is covered with soil, with now and then, on its top and sides, an exposure of rock. As we pass from its eastern extremity, along the top of it, we find no other rock than Magnesian Limestone, in place, until within

300 yards of the range of the shafts, where Sandstone is found both on the top and sides. At the eastern extremity of the ridge, the Magnesian Limestone is almost perfectly horizontal, with no perceptible dip, until it approaches the Sandstone, when it is seen dipping down. for a short distance, at an angle of ten or fifteen degrees, to the west. This Sandstone continues west for about 600 feet, visible at points, both on the top and sides of the ridge; but no other rock was seen (excepting on the south side, and near its base, where Dr. Shumard measured a brecciated mass, eighteen feet high, consisting of chert and Magnesian Limestone, until passing a short distance west of the range of shafts, where the Magnesian Limestone was again visible, with, at first, a dip of ten or fifteen degrees to the east; but a short distance further west, on the same ridge, it was horizontal. In the Sandstone, whether exposed on the ridge, or examined in the driftings, I found no appearance of stratification. The surface of the ridge is so covered with soil, that it is impossible to examine the eastern and western junctions of the Sandstone, with the Magnesian Limestone; but I infer, from examination of the driftings in the mine, that the western junction is irregular, with a general course across the ridge of about N., 20° W.; and that along this line, there is, in all probability, a space, for some distance beneath the surface, filled with the debris of the two rocks.

"Most of the mining done has been in a space, irregular, so far as explorations have shown, the direction of which is across and extending below the base of the ridge, and with a general course of about north, twenty degrees west, and of an estimated width of from forty to sixty feet, bounded on the east by Sandstone, and on the west by Magnesian Limestone. This, so far as explored, is found filled with tumbling rock, clay, chert, calc spar (semi-crystalline, and colored red by peroxide of iron), masses of iron ore and copper ores.

"From the vertical section, it will be seen that there are five shafts, the deepest of which (engine shaft) is 115 feet, and in which is the pump, worked by a steam engine. In sinking it, tumbling rock, of a magnesian character, was found through its entire depth, and it is cribbed from top to bottom. This shaft is connected by a level, 145 feet in length, with a shaft ninety feet deep, north of it, and thus connecting with the main works in the north hill. At the time of my last visit this level extended no farther than this shaft; but, since then, it has been run northwardly, the depth of about fifty feet below all preceding driftings; and, as I am informed by one of the company, with good success and fine prospects. Levels have been run into the north hill, from both the north and south sides; but most of the driftings have been fifty feet below these, and above which driftings, only (as represented on the vertical section), the ground has been stoped away.

"The copper ores found here are a mixture of the gray sulphuret and the green carbonate. Two analyses of a specimen, which was richer than the average run, gave the following results:—

	I.	II.
Silica,	1·16	1·29
Sulphur,	2·02	2·10
Peroxide of iron,	12·85	12·20
Oxide of copper,	61·16	60·16
Carbonic acid, water and loss,	22·81	24·25

Giving, as the mean of the two determinations, 48·41 per cent. of copper.

"The furnace for smelting the ore is distant from the mine about one mile, where there is an abundance of water during the whole year, for washing the ores, and supplying a blast for the furnace during eight months in the year. For this last purpose, however, the company have lately erected, at this point, a steam engine, and are now enabled to continue, at all seasons, their smelting operations. They are now engaged in smelting a large quantity of copper ore that has accumulated during the present year, and which, it is estimated, will produce thirty tons of copper; that, added to the twenty or thirty tons previously made, will make the total amount of copper made here, since the commencement of operations, in 1851, about fifty tons.

"During the first year of the operations of the Company, there was but little mining, most of the labor having been expended in erecting the furnace; and the average number of hands was not over six. During 1852, the average number of hands was about ten; and, at present, there are, probably, twenty or twenty-five in the employ of the company."

The owners deserve great credit for the energy with which they prosecuted the exploration of this mine, to prove the character of this and other copper deposits in the State.

TABLE OF LOCALITIES OF COPPER ORE.

The star (*) denotes the localities where the copper has been worked.

No. of Locality.	Name of Mine.	Town-ship.	Range.	Section.	County.	Whether on R.R. Land.	Kind of Ore.	Whether worked.	By whom reported.	Remarks.
1	....	35	2 E	....	Washington	....	Carbonate and Sulphuret	......	Litton	5 m's east of Caledonia.
2	Jordan	36	3 & 4	1 & 6	Washington	....	....	......	Litton	
3	Stanton	40	2 W	2	Franklin	....	Gray Sulphuret & green Carbonate	*	Litton	
4	Silver Hollow	40	1 W	N.W. 8	Franklin	....	Gray Sulphuret & green Carbonate	not now	Litton	
5	Hurst	41	1 W	26 & 27	Franklin	....	....	......	Shumard	
6	....	41	1	N.E. 28	Franklin	....	....	......	Shumard	
7	Bredell	41	1	34	Franklin	....	....	......	Shumard	
8	Copper Hill	40	2 W	24	Crawford	....	Gray Sulphuret, red Oxide and Blue Carbonate,	not now	Litton	R.R. land in this section.
9	....	38	3 W	14	Crawford	....	....	......	Shumard	
10	Reeves	39	3	13	Crawford	....	....	......	Shumard	
11	Bleeding Hill	38	2	4	Crawford	....	Carbonate, Oxide, Sulphuret and Native,	not now	Englemann	See Dr. King's Report.
12	& Hinch's	38	2	....	Crawford	....	....	......	Engelmann	
13	....	36	4	22	Crawford	....	....	......	Shumard	
14	Hibler	40	2	22	Crawford	....	Blue and Green Carbonate	*	Shumard	
15	....	36	5	15	Crawford	....	....	......	Shumard	
16	....	40	9 W	2	Maries	....	....	......	Broadhead	
17	....	30	24 W	19	Green	....	Sulphuret	......	Broadhead	In Lower Encr. Limest.
18	Haralson	29	24	10	Green	R	Sulphuret and green Carbonate	not now	Broadhead	Some found on RR. land.
19	....	29	25	2	Lawrence	....	Sulphuret	......	Broadhead	In Lower Encr. Limest.
20	Jos. Stogdill	30	25	S.E. N.E. 23	Dade	....	Sulphuret	......	Broadhead	In F. f. and i.
21	Jos. Stogdill	30	25	S.W. N.W. 24	Dade	....	Sulphuret and green Carbonate	......	Broadhead	In Lower Encr. Limest.
22	....	29	24	2	Green	....	Sulphuret	......	Swallow	
23	....	....	....	....	Dallas	....	Sulphuret	......	Swallow	
24	Goose	26	19	S.E. 9	Taney	....	Sulphuret and Carbonate	......	Swallow	In Magn. Limestone.
25	....	....	....	....	Benton	....	Sulphuret	......	Swallow	
							COPPER FURNACES.			
1	Stanton	40	2 W	N.E. $\frac{1}{4}$ 1	Franklin	....	Copper	not	Litton	Since 1851. Steam power.

ZINC.

Sulphuret of Zinc is very abundant in nearly all the mines in South-western Missouri, particularly in those mines in Newton and Jasper, in the Mountain Limestone. The Carbonate and the Silicate occur in the same localities, though in much smaller quantities. The ores of Zinc are also found in greater or less abundance in all the counties on the South-western Branch; but the distance from market, and the difficulties in smelting the most abundant of these ores, the Sulphuret, have prevented the miners from appreciating its real value.

It often occurs in such large masses as to impede very materially the progress of mining operations. For this reason, *Black Jack* is no favorite with the miners of the South-west. Many thousand tons have been cast aside with the rubbish as so much worthless matter; but the completion of the South-western Branch will so lessen the cost of transportation, as to give a market value to this ore, and convert into valuable merchandise the vast quantities of it, which could be so easily obtained in Jasper, Newton, and other counties of the South-west.

COAL.

There are but few localities of this valuable mineral west of St. Louis county; and those which do occur, are but unimportant outliers, very limited in extent and of ordinary quality. The beds in St. Louis county are extensive and very valuable. At its western terminus, this road again approaches the great western coal field, whose south-eastern boundary passes down through Vernon and Barton counties, into the territory west of Jasper, and thence south-west into Arkansas.

The western terminus, therefore, of this road, like the eastern, is near very extensive and valuable coal beds, over which the road must pass, if continued westward, beyond our State. Coal must ever be abundant and cheap on a road whose extremities are so near two inexhaustible coal fields.

LOCALITIES OF COAL.

Name of Mine.	Township.	Range.	Section.	County.	By whom reported.	Remarks.
....	45	6 & 7 E.	......	St. Louis.	Shumard	Many of the townships within this county are underlaid by the Coal Measures. Many mines are worked.
....	47	6 & 7	......	St. Louis.	Shumard	
Samuel Massey	36	4 W.	S.E. 21	Crawford	Shumard	Not now worked. Impure coal that contains too much Iron pyrites.
....	36	4	S.E. 30	Crawford	Shumard	Contains too much Iron pyrites.
....	38	6 W.	S. 9	Phelps...	Shumard	Cannel coal. Too much pyrites.

TIMBER.

The broad, rich bottoms of all the streams on the line of this road, sustain a very heavy growth of most excellent timber of nearly all the most useful varieties. Bur, red, laurel, pin, and swamp white oaks; black and white walnuts; white, blue and black ashes; white, red and wahoo elms; red birch, honey locust, buck-eye, box elder, black cherry, hackberry, pignut and common and thick shellbark hickories, red bud, sugar and white maples, mulberry, American plum, hazle, pawpaw, sycamore, muscadine, summer and fox grapes, and several species of thorn and willow, are most abundant.

The slopes and some of the high lands are covered with heavy forests of nearly all the trees found in the bottoms; while other portions of the high lands produce a medium growth of white, black, Spanish, post and chestnut oaks, shellbark hickory, sumachs, hazles and grapes. But a still larger part is sparsely timbered with small black-jacks, post oaks, and black hickories, forming the beautiful *oak-openings* of the south-west. This stunted growth is not, however, due to the poverty of the soil, but to the fires which have annually overrun this country since the earliest dates of the Indian traditions. These fires, fed by the rank annual growth of grasses and other herbaceous plants, have entirely destroyed some of the young trees, while they have scorched and very much retarded the growth of those sufficiently vigorous to withstand their ravages.

Large areas, particularly those underlaid by sandstones, are covered by very extensive and valuable forests of the yellow pine. These pine forests are very extensive in McDonald.

The spontaneous growth of the Osage-orange in the southwest, proves its adaptation to the climate and soil, and indicates its fitness for hedges in that region.

ABSTRACTS FROM DR. B. F. SHUMARD'S REPORT.

"*Crawford County.*—The valleys of the larger streams are frequently heavily timbered with white and bur oak, white and black walnut, white and sugar maple, shellbark hickory, pawpaw, dogwood, linden or basswood, grape and haw. On the higher uplands, between the rough ridges skirting the streams, we frequently find extensive tracts of level, post and black oak, and hickory lands. In Sec. 16, T. 36, R. 4 W., there is a *pinery.*

"*Phelps County.*—The valleys of the larger streams, Little Piney, Spring creek, Dry Fork of the Meramec and Bourbeuse, are in general very heavily timbered with white and bur oak, shellbark and pignut hickory, white and black walnut, sugar and white-leafed maple, dogwood, linden, hackberry, honey locust, cottonwood, thorn, and several varieties of grape.

"*Pulaski County.*—The valleys of the Gasconade, Big Piney river, and Robideaux and Spring creek, sustain a heavy growth of white, bur and scarlet oak, white and sugar maple, shellbark hickory, white and slippery elm, dogwood, cottonwood, ash, linden, elder, grape, hackberry, and white and black walnut. The hills, also, in the vicinity of the larger streams are heavily timbered.

"*La Clede County.*—The bottoms of the larger streams support a heavy growth of the finest kind of timber. On the upland, "post oak flats" we find post and white oak, and blackjack and black hickory."

ABSTRACTS FROM MR. BROADHEAD'S NOTES.

"The hills near the Gasconade river are well timbered, also between Little Tavern and Sugar creek, and between Dry creek and Clifty creek, consisting mostly of white and post oak, and black oak. There is also a fine timbered tract adjacent to Lane's prairie, and we there find laurel oak, shellbark and pignut hickories, mulberry, black walnut, red and white elm, plum, sassafras, ash, and grape vines.

"But the best timbered lands we find along the larger streams, including the Gasconade with Big and Little Maries, and Dry Fork of Bourbeuse, where we find bur, laurel, red, rock chestnut, swamp white oak, with shellbark and pignut hickory, hackberry, black and white walnut, American and red elm, sycamore, linden, red bud, pawpaw, and grape.

"There is not so much nor such fine timber on the smaller streams (Little Tavern, Spring creek, Sugar creek, Cave Spring and Dry creek), but there is still some fine timber on them. We find pignut hickory, chinquepin oak, rock chestnut oak, with sometimes laurel and white oak; also hazle, American elm, red elm, alder, ironwood, hornbeam, red bud, pawpaw, and the muscadine grape frequently abounds; found the muscadine on Dry creek and the smaller streams flowing into the Gasconade. On Cave Spring creek and Dry creek found gum trees."

In short, the timber of this part of the State is good and sufficiently abundant to supply all the demands of a dense and industrious population. The various kinds of trees and shrubs observed, are shown by the following catalogue.

TREES AND SHRUBS.

ALDER.

Common Alder (*Alnus serulata*). On streams in Newton, Lawrence and Taney counties; also on the tributaries of the Gasconade river, in Maries county.

Black Alder or *Winter-Berry*. In wet land and wooded bottoms.

APPLE.

Crab Apple (*Malus coronaria*). Bordering rich prairies.

ASH.

White Ash (*Fraxinus Americana*). Abundant near Lane's prairie, and on Bourbeuse creek, in Maries county.

Blue Ash (*Fraxinus quadrangulata*). On good soil.

Prickly Ash (*Zanthoxylum Americanum*). In bottoms and moist places.

BASSWOOD OR LINDEN.

American Linden or *Lime* (*Tilia Americana*). On Sac and Gasconade rivers. On good, rich soil.

BIRCH.

Red Birch or *River Birch* (*Betula rubra*). On borders of nearly all the streams.

BLACKBERRY.

Low Blackberry or *Dewberry* (*Rubus Canadensis*). In open forests.

Wedge-leaved Blackberry (*Rubus cuneifolius*). In forests adjacent to the bottoms of all the larger streams.

BLADDER-NUT.

American Bladder-nut (*Staphylea trifolia*). In rich bottoms and on the debris at the bases of the bluffs.

BUCKEYE.

Large Buckeye (*Æsculus lutea*), in rich bottoms.

BLUEBERRY.

Huckleberry (*Vaccinium*). On flint hills, in Taney, Green, Maries and Gasconade counties.

Blueberry (*Vaccininm vacillans*). On flint hills, in Taney and Green counties.

BOX ELDER.

Box Elder or *Ash-leaved Maple* (*Negundo aceroides*). Abundant in rich bottoms.

BURNING BUSH.

Burning Bush (*Euonymus atropurpureus*). On Little Pomme de Terre—very beautiful when in fruit.

BUTTON-WOOD.

Sycamore (*Platanus occidentalis*). In the bottoms of all our principal streams.

BUTTON BUSH.

Button Bush (*Cephalanthus occidentalis*). In wet places and beside streams—not very abundant.

CEDAR.

Red Cedar (*Juniperus Virginiana*). On dry limestone bluffs, along many of the larger streams. Very abundant in Franklin county, near the Meramec river.

CHERRY.

Black or *Wild Cherry* (*Cerasus serotina*, D. C.). On the best soils.

COFFEE TREE.

Coffee Tree (*Gymnocladus Canadensis*). In rich soil, bottoms and highland.

COTTON-WOOD.

Cotton-wood (*Populus Canadensis*). On river bottoms; not very abundant in south-west Missouri.

CORAL BERRY.

Coral Berry or *Indian Currant* (*Symphoricarpus vulgaris*). Everywhere on good soil.

CURRANT.

Currant (*Ribes*). Several species, but none are abundant.

DOGWOOD.

Flowering Dogwood (*Cornus Florida*). On bluffs and ridges; generally very sparse, but found in most of the counties.

Panicled Dogwood (*Cornus paniculata*). In ravines and wet bottoms.

Rough-leaved Dogwood (*Cornus asperifolia*). Not abundant.

Silky Dogwood (*Cornus cericea*). In wet bottoms.

ELDER.

Common Elder (*Sambucus Canadensis*). Very large in the rich bottoms.

ELM.

White or *American Elm* (*Ulmus Americana*). Found on the richest soils, in all the counties.

Slippery Elm or *Red Elm* (*Ulmus rubra*). On good soils, in all the counties.

Wahoo Elm (*Ulmus alata*). In Green county, on limestone ridges—scarce.

GRAPE.

Summer Grape (*Vitis æstivalis*). Abundant on good soils.

Fox Grape (*Vitis labrusca*). On good soil.

Winter or *Frost Grape* (*Vitis cordifolia*). On good soil.

Muscadine (*Vitis vulpina*). On rocky ridges and rich bottoms.

River Grape (*Vitis riparia*). In alluvial bottoms.

GREEN BRIER.

Green Brier (*Smilax rotundifolia*). In thickets and beside fields.

Glaucus Green Brier (*Smilax glauca*). In thickets, ravines, and beside roads.

Smilax pseudo-China. In Taney county, along the richer valleys.

Smilax bona-nox. Abundant on shrubs by the fences.

Smilax quadrangularis. In thickets and fence corners.

Smilax hastata. In thickets on good soil.

GOOSEBERRY.

Prickly Gooseberry (*Ribes cynosbati*). Abundant.

Wild Gooseberry (*Ribes rotundifolia*). In woods and on borders of prairies. On rich land.

GUM.

Black Gum (*Nyssa multiflora*). In Maries county.

HACKBERRY.

American Nettle Tree or *Hackberry.*

(*Celtis occidentalis*). In rich soil.

Hackberry (*Celtis incrassifolia*). In rich soils and low grounds.

HAZLE.

American Hazle (*Corylus Americana*). In rich prairies and on the borders of the forests.

HAW.

Black Haw (*Viburnum prunifolium*). In open forests, on good soil.

Red Haw (see Thorn).

HICKORY.

Common or *Mockernut Hickory* (*Carya tomentosa*). In dry, good soil.

Shellbark Hickory (*Carya alba*). Not common.

Thick Shellbark Hickory (*Carya sulcata*). Only in very rich bottoms.

Pignut Hickory (*Carya porcina*). Rare on good soil.

Black or *Bullnut Hickory* (*Carya microcarpa?*). Very abundant on poor soil, associated with black-jack and post oak.

Bitternut Hickory (*Carya amara*). On Caps creek, in Newton county.

HONEYSUCKLE.

Yellow Honeysuckle (*Lonicera flava*). In Taney and Green counties.

Small-flowered Honeysuckle (*Lonicera parviflora*). Rare.

HORNBEAM.

Hop Hornbeam (*Ostrya Virginica*). Near streams and rocky branches on hill-sides.

American Hornbeam or *Ironwood* (*Carpinus Americana*). On rocky hill-sides and sometimes in the bottoms of the smaller creeks.

IRONWOOD (see Hornbeam).

JUDAS TREE.

Red Bud or *Judas Tree* (*Cercis Canadensis*). Abundant on good soil.

LOCUST.

Sweet or *Honey Locust* (*Gleditschia triacanthos*). In the richest soils. Not so abundant as in some other parts of the State.

LINDEN (see Basswood).

MAPLE.

White Maple (*Acer eriocarpum*). In the river bottoms, on sandy alluvium.

Sugar Tree (*Acer nigrum?*). In Taney, Green, &c.

MULBERRY.

Red Mulberry (*Morus rubra*). On rich lands; generally scarce.

NETTLE TREE (see Hackberry).

OSAGE-ORANGE.

Osage-orange (*Maclura aurantiaca*). In the valley of Spring river.

OAKS.

First Division—Leaves lobed, lobes rounded.

White Oak (*Quercus alba*). Dry soil, generally on hills; but is often found in the valleys of the lesser streams. On all the better lands in Maries county, valleys and ridges in Taney, and on the high lands of medium quality in all the counties.

Overcup White Oak or *Bur Oak* (*Quercus macrocarpa*). On rich soils—good timber.

Post Oak (*Quercus obtusiloba*). Dry, poor soils; timber most durable of all our oaks; very common.

Second Division—Leaves coarsely toothed.

Swamp White Oak, often called *Bur Oak* (*Quercus bicolor*). On low, rich and damp soil.

Chestnut White Oak (*Quercus prinus*). Wet, rich soil, in shaded places.

Rock Chestnut Oak (*Quercus monticola*). Dry soil, on rocky bluffs and ridges.

Chestnut or *Yellow Oak* (*Quercus acuminata*). On limestone bluffs and dry bottoms.

Chinquapin or *Dwarf Chestnut Oak* (*Quercus prinoides?*). On limestone bluffs larger than usual. Acorns often peduncled.

Third Division—Leaves entire.

Laurel Oak, erroneously called *Pin Oak* (*Quercus imbricaria*). On the borders of prairies and fields. On rich land.

Fourth Division—Leaves lobed, lobes mucronate.

Black Jack Oak (*Quercus nigra*). On the poorest soils; very common on the barrens in all the counties of South-west Missouri.

Black Oak (*Quercus tinctoria*). Abundant on good and medium soil—excellent timber.

Scarlet Oak (*Quercus coccinea*). On good soil.

Red Oak (*Quercus rubra*). On damp, rich soil, especially in the bottoms.

Pin Oak (*Quercus palustris*). In swamps and wet land—scarce in South-west Missouri.

Gray Oak (*Quercus ambigua?*). On good soil.

Spanish Oak (*Quercus falcata*). Rare in the South-west.

PAWPAW.

Pawpaw (*Anona triloba*). In rich soils, particularly under limestone bluffs.

PERSIMMON.

Persimmon (*Diospyros Virginiana*). Scarce in South-west Missouri. In good soil; borders of prairies and fields.

PINE.

Yellow Pine (*Pinus mitis*). In Crawford, McDonald, and Washington counties.

PLUM.

Red Plum (*Prunus Americana*). In bottoms, and on borders of the prairies—rich soil.

Chickasaw Plum (*Prunus Chicasa*).

PRICKLY ASH.

Prickly Ash (*Zanthoxylum Americanum*). In wet places, on the borders of prairies and forests.

RATAN.

Ratan (*Wistaria frutescens*).

ROSE.

Prairie Rose (*Rosa setigera*). Prairies and open forests. Several other species were observed, as *R. blanda* and *R. lucida.*

RASPBERRY.

Red Raspberry (*Rubus strigosus*). On borders of fields and forests.

Black Raspberry or *Thimbleberry* (*Rubus occidentalis*). In open forests and beside roads and fields.

SYCAMORE.

Buttonwood or *American Plane Tree* (*Platanus occidentalis*). In the bottoms of all the principal streams.

SUMACH.

Dwarf Sumach (*Rhus copallina*). Common by the borders of fields, roads and prairies.

Smooth Sumach (*Rhus glabra*). Road sides and open forests.

Staghorn Sumach (*Rhus typhina*). Often in clusters in prairies.

Poison Ivy or *Poison Oak* (*Rhus toxicodendron*). On rich soils, large and abundant.

Fragrant Sumach (*Rhus aromatica*). Abundant in forests and by roads and fields.

SPIRÆA.

Flowering Spiræa, Nine-bark or *Seven-bark* (*Spiræa opulifolia L.*). On limestone bluffs, bordering streams.

Spiræa corymbosa. On dry prairies and ridges.

SASSAFRAS.

Sassafras (*Laurus Sassafras*). On medium soil; not common.

SARSAPARILLA.

Sarsaparilla. Near streams, on rich soil.

SERVICE-BERRY.

Wild Service-Berry or *Shad-bush* (*Amelanchier Canadensis*). On bluffs and in forests. Common on most streams, most abundant on Pomme de Terre and Swan creek.

STAFF-TREE.

Staff-Tree (*Celastrus scandens*). On river banks and broken bluffs.

THORN.

Black Thorn (*Cratægus tomentosa*). In rich forests.

Red Haw (*Cratægus coccinea*). Abundant in open forests.

Dotted Thorn (*Cratægus punctata*). On bluffs and ridges.

The following species of Thorn were also observed:—*Cratægus crus-galli*, *Cratægus spathulata*, *Cratægus apifolia*.

TRUMPET-CREEPER.

Trumpet Creeper (*Tecoma radicans*). In most counties on rich soil, climbing over trees.

Virginian Creeper (*Ampelopsis quinquefolia*).

WALNUT.

Black Walnut (*Juglans nigra*). In bottoms and common on high, rich soil. In all the counties.

White Walnut or *Butternut* (*Juglans cathartica*). In low, rich soil, and under bluffs.

WILLOW.

Salix. Several species were observed on the borders of the several streams.

WINTER-BERRY.

Winter-Berry (*Prinos lævigatus*). In low, wet forests and thickets.

WITCH-HAZLE.

Witch-Hazle (*Hamamelis Virginica*). Taney county, on Swan creek.

WATER.

There is, probably, no part of this continent that can boast of so large a number of bold, limpid springs, whose pure, cool waters gush forth in such abundance to beautify and refresh the land. Bryce's Spring, on the Niangua, is one of the largest. It rises in a secluded valley where it forms a small pond and then flows away—a *river*. This river, just below where it flows from the spring, is 126 feet wide, and has an average depth of about one foot, and its velocity is a little more than one foot per second. This immense spring discharges more than 126 cubic feet of water per second, 455,326 per hour, and 10,927,872 cubic feet per day. The water is nearly pure, sustains about the same temperature at all seasons, and shows no perceptible fluctuations in quantity either in the dryest or wettest seasons.

This is one of the many hundred large springs whose pure waters unite and form the numerous streams which flow from this table land. Many of them furnish the very best water-power for driving mills and factories. In some respects it is more desirable than that offered by ordinary streams.

The water is so warm during the winter that no ice forms about the wheels or other machinery.

The supply of water is constant, and the quantity about the same at all seasons, so that the works are never endangered by freshets, or compelled to remain idle for the want of the usual quantity of water.

These advantages have been fully tested and are duly appreciated, as one would judge from the great number of mills located on these springs.

The streams formed by these springs are numerous, clear and rapid, furnishing sufficient water-power to drive all the mills and factories demanded by any ordinary population. While the springs and streams large enough to furnish good mill sites are very numerous, the smaller fountains and branches are so abundant, that every farm may be supplied. In short, the pure, limpid fountains and streams of this region are unrivalled in beauty and adaptation to the wants of man—they have challenged the admiration and praise of every traveler.

POPULATION.

So little has been known of the vast resources and numerous advantages of this region, that many parts of it are only sparsely populated. But a hardy, energetic, intelligent and thrifty yeomanry are rapidly opening the country, while the flourishing towns are filling up with a wealthy and refined population.

SCHOOLS AND CHURCHES.

Public schools are everywhere established, and most liberally sustained by donations of lands and a large part of the State revenues. Every town has its private schools established by the munificence of the citizens, and sustained by a universal desire for a complete and thorough education. Many of them have fine buildings, and are most excellent schools.

Churches are numerous, varying in style from the plain log building to the elegant brick edifice. Many of them are most tastefully located amid the beautiful scenery of this favored country.

I have thus attempted, in a somewhat hasty and imperfect manner, to delineate some of the more important natural resources and advantages of the county through which the South-western Branch of your road has been located and partially built. Great care has been taken to keep within the bounds of facts well authenticated and such deductions as might be legitimately drawn from those facts. It may be proper here to state, that our explorations have been, of necessity, but partial, and there can be no doubt that a full and careful examination of this county would enable us to more than double the very extensive catalogue of mineral localities already known.

COMPANY LANDS.

The lands of your company contain more than a fair proportion of the good soil and mineral wealth of this region. The town of Granby and other localities of lead, which appear equally good, and many of the best iron beds, are on the lands donated to the South-western Branch. The value of your mineral lands is very great, and can scarcely be realized until the road is completed and the mineral raised will sell for something near its market value in other localities. The good agricultural lands cover a vast extent of fine country; and their value, already somewhat appreciated, will be greatly increased by the completion of the road. All of the poorer lands will be demanded for timber and pasture; and the day is not far distant when those *broken ridges* and *flint hills*, which have heretofore been deemed worthless, will command the highest prices for the cultivation of the grape.

Some of your mineral lands would be cheap at $1,000 per acre, and $10 per acre would be a very low valuation for the whole when the road is completed. At this minimum value they would yield more than $11,000,000.

We have thus presented, in as brief a space as possible, some of the leading features of the country through which the South-western Branch of the Pacific Railroad passes. The facts presented will doubtless cause all to appreciate in some good degree the numerous and distinguished natural advantages of this favored portion of our State; but to realize the whole truth, one must see for himself; he must inhale the pure bracing air beside the bold fountains and limpid streams, on the broad waving prairies and in the extensive oak openings; he must descend into the cavern and the mine, and behold the glowing furnaces sending forth their streams of iron and lead; he should examine the broad acres ripe for the golden harvest, and enjoy the generous hospitality of the country farm-house and the village mansion;—in short, he should see for himself this favored country, possessed by a hardy, energetic and noble people.

APPENDIX.

(A.)

THE CASSWELL MINE.

Since the foregoing report was written, I have had the pleas ure of visiting this valuable mine in company with the Hon. John F. Darby, one of the proprietors. It is situated on the N. ½ of the S.W. qr. Sec. 34, T. 42, R. 1 E., in the bluff of the southern side of the Meramec. At the mine the bluff rises rather abruptly to the height of some 200 feet. The rocks at the base are the upper beds of the 3d Magnesian Limestone, and those cropping out on the brow of the hill, near the top, are the lower beds of the 2d Sandstone.

The vein was discovered in the fall of 1855 by Mr. Brewer, who opened the mine and raised about 12,000 pounds of galena. It was next worked by Mr. Erie Standifer, who took out some 15,000 pounds of the ore. Mr. Michael Dolan has worked it from time to time since 1856 under the direction of the present owners, Messrs. Darby, Vandeventer and Beardslee, and has raised about 100,000 pounds of good galena.

Mr. Dolan's systematic operations very clearly indicate the characters and value of the vein. It cuts through the bluff in a direction nearly north and south, and almost perpendicular, but inclining a little to the east in its descent. The lead ore is nearly all the sulphuret, though the carbonate sometimes occurs. The gangue is heavy spar, calc spar, and red clay. The thickness of the vein varies from two inches to ten. This vein, like the Evans, cuts through the lower beds of the 2d Sandstone and down into the 3d Magnesian Limestone, which forms the base of the bluff. From a point on the slope near the base of the Sandstone a shaft has been sunk 95 feet into the Limestone, and an adit has been run on the vein some 200 feet from near the base of the bluff, intersecting the shaft above named. The appearance, position and direction of this vein seem to indicate that it is a continuation of the Evans Lode,* on the south side of the ridge, which some have supposed to be a part of the Mount Hope vein.

Whether these veins shall prove to be one and the same, and whether they prove to be *true veins*, extending down indefinitely, or merely to the base of the formation in which they are found, they can not fail to be extensive and valuable. The length of the two is not less than one mile, and the average depth of the parts not worked, to the bottom of the 3d Magnesian Limestone, can not be less than 300 feet, and is probably between 400 and 500 feet.

In estimating the profits of mining on these veins, it will be safe to put down the length at one mile and the depth below the Sandstone at 400 feet, and that the remainder of the vein will prove as rich or even richer than the parts worked out. But these estimates are made upon the most unfavorable opinions respecting the character of these lead veins. The opinion expressed by some geologists that these are only *Gash-veins*, and confined to one formation, the 3d Magnesian Limestone, has no support in the appearance of the country or the character of the veins themselves. And I submit the proposition, with all due deference to the opinions of others, that no geologist can examine the phenomena presented by this vein, and the Evans Mine, and the Virginia Mine, and make them conform in any tolerable degree to the definition given of a *Gash-vein*. On the contrary, all the facts observed point most significantly to the characters of *true veins—veins* which extend downwards indefinitely, without regard to the limits of formations. With this view of the character of these veins, which I conceive to be the true one, the value of these and the neigh-

* See page 49.

boring mines will be vastly increased, as there will be no fear of exhausting them.

The unfavorable opinion respecting the Lead Mines of Missouri, which has prevailed to some extent among foreign miners and capitalists, has arisen, I apprehend, from the erroneous opinions of some geologists that our mines have characteristics and geological relations similar to those of the Wisconsin mines. While I shall not deny that some of our lead veins resemble those of Wisconsin, and appear like *Gash-veins*, there are many others in which the analogy does not hold good in any one important character.

1. According to Mr. Whitney, the valuable Lead veins of Wisconsin are confined to a formation not more than 100 feet thick; but in Missouri the most valuable veins range through three formations, the aggregate thickness of which is not less than 1,000 feet.

2. In Wisconsin the Lead veins are limited to one formation in the upper part of the Lower Silurian System, while in Missouri the most valuable veins range through two members of the Carboniferous system and the two lower formations of the Silurian.

3. While in Wisconsin, so far as I know, there are no evidences of extensive igneous action or violent disturbances in the neighborhood of the lead mines, in Missouri, both within and around the lead field, there are most decisive proofs of extensive igneous action and violent disturbances—mountains of granite and porphyry have been thrown up—mountains and ridges of porphyry have been fractured and rent asunder, and the fissures filled with dykes (veins) of granite, greenstone, quartz, basalt, dolerite and porphyry, and *true veins* of copper and wolfram,* and veins (where the metalic ore fills the entire fissure) of specular iron and galena; some of these dykes pass into the sedimentary rocks, changing the sandstone to quartzite and the limestone (the lower lead-bearing beds) into crystalline marble.

4. In Wisconsin the profitable veins have not extended more

* Dr. Norwood is my authority for these veins of copper, wolfram and dolerite.

than 100 feet in depth, but in Missouri two shafts have been sunk on the large Virginia vein to the depth of 260 feet, without any diminution or indication that it would run out.

5. In Missouri some of the veins do pass from the Limestone into the Sandstone above, as seen at the Evans and the Caswell Mines.

6. Many of the veins in Missouri present all the appearances of *true veins*; dislocations and disturbances have been produced by powerful agencies, as indicated in some places by the fragments of the original strata filling a part of the fissure, by well marked and extensive slickensides, by the displacement of the strata, and the irregularity of the fissure.

7. The veins are often very long; some have been explored more than one mile.

8. In many mines the fissures are filled as they usually are in *true veins;* the sheet of galena runs through the middle, with a gangue of heavy spar or calc spar, or both, on each side.

9. Selvages, so remarkable in *true veins*, also occur in the Missouri mines.

Such are some of the facts which should lead us to suspect the validity of all arguments drawn from any apparent analogy between the Wisconsin mines and our own. And besides, even on the supposition that our veins do not extend below the base of the 3d Magnesian limestone, there is still from 200 to 400 feet of this rock below the deepest workings of nearly all the mines in the counties of Jefferson, Franklin, Crawford, and the north of Washington; while in the South-west the lead-bearing portion of the Mountain Limestone is at least 200 feet thick. Below these beds are the Chemung rocks, which are not over 100 feet in thickness. Whether the lead passes down through this formation is not known, as no vein has been traced or worked to it. The character of the rock, however, does not indicate the existence of valuable veins, though some deposits of lead and copper have been discovered in it. In passing from the Chemung rocks near the northern boundary of Taney, we come directly upon the lead-bearing rocks of that county, which are the 2d and 3d Magnesian Limestones. The 1st and 2d Sandstones are very thin or entirely wanting in this part of the county, while the 2d and 3d Magnesian Limestones present an aggregate thickness varying from 600 to 1,000 feet.

These facts show that the mines of Newton and Jasper have beneath them at least 1,000 feet of lead-bearing limestones, and those in Taney from 600 to 800 feet of the lower part of the same beds. In view of these conclusions, based as they are upon the most unfavorable opinions entertained by any of the character of our veins, the miner and capitalist need not fear the exhaustion of our lead mines; but, when they take into consideration the facts above stated, which show an entire want of analogy between our own mines and those on the Upper Mississippi, and which point so conclusively to the most reliable characteristics of true veins, their fears, if any still exist, that our lead mines have seen their most prosperous days, must be banished, and they will continue their operations with brighter hopes of eminent success.

These views are fully sustained by the most recent developments of our mines, as some of the oldest have been reopened and worked with greater success than ever before; and besides the deepest diggings have often proved the most profitable.

While, then, it may be true that some of our lead deposits are only *gash-veins*, others (and among them the Casswell) give every indication of being *true veins*.

(B.)

GRANBY LEAD MINES.

It has been somewhat difficult to get a reliable statement of the amount of lead made from the Granby Mines; but from the best information derived from the miners, the smelters, and the present proprietors, Messrs. Blow & Kennett, the amount of lead made can not be less than seven million or eight million pounds. I am indebted to Mr. Blow for the following statement of the operations of the company:

"From the 1st February to 1st September, 20,000 pigs, averaging 80½ lbs., have been smelted at Blow & Kennett's furnace at Granby; the transportation of which—by wagons, railroads, and steamboats—to St. Louis, has amounted to nearly twenty thousand dollars.

"Five steam engines and over two hundred men are employed by this firm in mining and smelting, while a much larger number are engaged in mining other shafts on the Granby section, all belonging to the Pacific Railroad, but under lease to Blow & Kennett.

"The usual cost of lead per 100 lbs., from the furnace to St. Louis, is $1.25. The usual price of mineral is $16 free of rent, or $14 and rent."

(C.)

PARK'S COPPER MINE.

I am indebted to Mr. J. V. Phillips for a full report, illustrated with numerous sections and maps, upon the Copper Mine of Mr. Andrew Park, in Sec. 17, T. 40, R. 1 E., Washington county. I regret that I can not publish it in full, as it would not be intelligible without the sections. From it I make the following summary and extracts:

"The vein is in the upper part part of the 3d Magnesian Limestone, and appears to run parallel with the strata, which dip about 10° toward the center of the ridge. It is seen on the sides of the ridge in several places for more than a mile in extent, and has been opened in three localities; in one, the level was extended fifty feet on the lode. The vein contains the green and blue carbonates and the yellow and gray sulphurets of copper, in a gangue of clay, heavy spar, calc spar and oxide of iron in cherty matter.

"Mr. Park, who had charge of the mining operations, thinks the vein showed a disposition to open out, about every eight feet, in vertical seams or crevices. These openings are filled with decomposed flint and ferruginous matter, and are about one foot wide.

"The ore in all the openings evidently belongs to the same vein, which is horizontal, and will doubtless follow the dip of the Limestone to the center of the ridge; and each ridge may be supposed to form a *copper basin*, and the central basin to be the center from which these ridges radiate. The richest portion of the vein or deposit may be looked for near the center of the basin. There is evidently a large amount of copper ore in these basins, and it lies in a good position for economical mining."

About ten tons of the ore has been taken out; it yields about twenty per cent of copper.

Mr. Phillips estimates the profits on every hundred tons of ore raised and shipped to Baltimore, at $5,950.

ERRATA.

On page 1, 5th line, for "have" read "has."

On page 8, 7th line from bottom, for "a part" read "parts."

On page 10, for the amount of potash in No. 12 A, for "6.3368" read "0.3368."

On page 48, 1st line, for "stopped" read "stoped."

www.ingramcontent.com/pod-product-compliance
Lightning Source LLC
LaVergne TN
LVHW021425110826
845150LV00007B/2087

* 9 7 8 1 4 2 5 5 0 8 1 2 8 *